CAHOON'S FORMULATING X-RAY TECHNIQUES

CAHOON'S FORMULATING X-RAY TECHNIQUES

ninth edition

Thomas T. Thompson, M.D.

Associate Chief of Staff, Education,
Durham Veterans Administration Hospital;
Associate Professor of Radiology,
Duke University Medical Center

Technical reviews of the text and editorial assistance by

John E. Cullinan, R.T. (F.A.S.R.T.), Eastman Kodak

Terry R. Eastman, R.T. (F.A.S.R.T.), Du Pont

Reid E. Kellogg, Ph.D., Du Pont

Lee Mockenhaupt, R.T., Minnesota Mining and Manufacturing

Duke University Press   Durham, North Carolina   1979

© 1979 by Duke University Press
LCC Number 79–87805
ISBN 0–8223–0431–7
Printed in the United States by
Kingsport Press, Inc.

To George F. Baylin, Professor of Radiology and R. J. Reynolds Professor of Medical Education, Duke University Medical Center, whose contributions to the education of all those with whom he comes in contact is legendary; and to Nancy, Mark, and Carol

CONTENTS

Contents

PREFACE TO THE NINTH EDITION

As before, the purpose of this text is to provide the radiologic technologist with the art and basic principles of formulating radiographic techniques, as well as a useful understanding of the scientific fundamentals that support these techniques. It is intended to be a compact yet comprehensive text on radiographic exposure, radiographic fundamentals, and technique conversions. It is a book concerned with radiographic techniques as they are applied to produce a diagnostic medical radiograph.

The organization of this edition is different from previous editions. Pertinent study and review questions have been added to each chapter. These questions have been designed in the same format as those used on the typical registry examinations. Much new material has been added including information on focal spots of X-ray tubes, automatic (mechanical) film processing, "rare earth" intensifying screens, xeroradiography, electron radiography, subtraction techniques, and technique conversion systems including the Du Pont Bit System and the Siemens Point System. The chapter on pathological conditions has been greatly expanded and new concepts on technique conversions for pathological conditions introduced. Most of the radiographs have been replaced by new illustrations produced by the LogEtronics method described in the book.

ACKNOWLEDGMENTS

In accepting the task of revising *Cahoon's Formulating X-ray Techniques* I felt that a ninth edition, updated and therefore more useful than the previous editions, would be a fitting tribute to the late John Cahoon and his contributions to the field of radiologic technology.

Toward this end the best authorities in the field were approached to review the eighth edition, suggest material and format for the ninth edition, and provide critiques of the ninth edition manuscript for technical accuracy and up-to-date presentation. Among these technical experts the following have been especially helpful: John E. Cullinan, R.T. (F.A.S.R.T.), of Eastman Kodak, Rochester, New York; Terry R. Eastman, R.T. (F.A.S.R.T.), of E. I. du Pont de Nemours, of Dallas, Texas; and Lee W. Mockenhaupt, R.T., of the Minnesota Mining and Manufacturing Company, Saint Paul, Minnesota. Dr. Reid E. Kellogg, Director of Technical Services of Du Pont in Chestnut Run, Delaware, has given me invaluable research data. Not only did the technical reviewers unselfishly contribute their time and effort towards this publication but their companies did as well. Many thanks go to these companies for their dedication to the furtherance of the education of radiologic technologists.

The manuscript benefited greatly because of a review of the pedagogy employed in the text as provided by Cindi Shuba, B.S., R.T., and the students of the radiologic technology programs of the Veterans Administration Hospital in Durham, North Carolina, and Duke University Medical Center.

The exhaustive review of the final draft was made by James Wasseen, M.S., R.T., Administrative Director of the Department of Radiology of the Medical College of Virginia.

Ronald Mitchell, Chief of the Medical Media Production Service at Duke University, and his staff provided immeasurable help and suggestions with the illustrations contained in this book.

The instructor's manual for the ninth edition was prepared by Joanne S. Greathouse, M.Ed., R.T.-R., Assistant Professor, Radiologic Technology, Medical College of Virginia.

Finally, this ninth edition could not have been possible without the help of those mentioned above or without the help of Ted Saros, Associate Director of Duke University Press. Moreover, the many hours spent over the typewriter by Ms. Cathy Collins and Ms. Elma C. Medley are greatly appreciated.

Thus, this textbook is the end result of the labor of many whose chief desire is that the student will be a better technologist because of the information contained in this book.

Thomas T. Thompson

CAHOON'S FORMULATING X-RAY TECHNIQUES

1 SIMPLE MATHEMATICS IN RADIOGRAPHY

A basic knowledge of simple mathematics is crucial for understanding and developing radiographic techniques. Many problems can be reduced to the application of simple mathematical equations. Technique problems must be stated and conversions must be calculated accurately. Calculation of exact technique factors represents a much more scientific approach than simply guessing what technique should be used for a particular examination. (Incidentally, guesses in matters of technique are almost invariably wrong.) This chapter is devoted to refreshing our memory of a few fundamentals of mathematics—fractions, decimals, percentages, equations, ratio and proportions, and the metric system.

FRACTIONS

A fraction is one or more of the equal parts into which a number may be divided. A simple fraction is expressed by writing two numbers separated by a line, e.g., $\frac{2}{3}$, $\frac{1}{4}$, $\frac{5}{6}$, or $\frac{2}{3}$, $\frac{1}{4}$, $\frac{5}{6}$. The number below or after the line is called the *denominator;* it represents how many parts a number or object is divided into. The number above the line is called the *numerator;* it represents how many parts of the whole number have been taken. For example, the fraction $\frac{3}{4}$ indicates that three parts (the numerator) of a whole which has been divided into fourths (the denominator) have been taken. The fraction sign—the bar "–" or the slash "/"—also indicates that the numerator is to be divided by the denominator. Thus $\frac{3}{4}$ indicates that 3 is to be divided by 4.

Multiplying fractions

To multiply fractions, the product of all the numerators is placed over the product of all the denominators.

Example:
$$\frac{1}{2} \times \frac{2}{3} \times \frac{5}{6} = \frac{1 \times 2 \times 5}{2 \times 3 \times 6} = \frac{10}{36}$$

The word "of," as used with fractions, means "times": $\frac{3}{4}$ of $\frac{1}{2}$ means $\frac{3}{4} \times \frac{1}{2}$.

Reducing or simplifying fractions

To reduce or simplify a fraction is to write the fraction with the lowest possible whole number denominator and numerator. Both must be acted upon at the same time and by the same divider. Thus, divide the numerator and denominator by a number that goes into both without remainder.

Example: Simplify $\dfrac{10}{36}$.

Solution: Neither the numerator nor the denominator can be divided without remainder by any numbers except 1 and 2. Dividing each by 1 would not simplify the fraction; it would remain $\frac{10}{36}$. However, dividing both numerator and denominator by 2 allows the fraction to be reduced to $\frac{5}{18}$:

$$\frac{10}{36} = \frac{2\,\overline{)10}}{2\,\overline{)36}} = \frac{5}{18}$$

Division of fractions

To divide one fraction by another, multiply the first fraction by the *reciprocal* of the second fraction. The reciprocal is the fraction turned upside down; e.g., the reciprocal of $\frac{1}{5}$ is $\frac{5}{1}$. Otherwise expressed, the reciprocal of a fraction is that number which when multiplied by the fraction yields the result 1:

$$\frac{1}{5} \times \frac{5}{1} = 1$$

Example: Divide $\frac{1}{4}$ by $\frac{1}{5}$.

Solution:

$$\frac{1}{4} \div \frac{1}{5} = \frac{\frac{1}{4}}{\frac{1}{5}} = \frac{1}{4} \times \frac{5}{1} = \frac{1 \times 5}{4 \times 1} = \frac{5}{4}$$

POWERS AND DECIMALS

Familiar examples of *powers* are squares—e.g., $3^2 = 3 \times 3$—and cubes— e.g., $3^3 = 3 \times 3 \times 3$. The power is expressed by the superscript, which indicates how many times a number is multiplied by itself. By convention, $3^1 = 3$; then $3^2 = 9$ and $3^3 = 27$, and so on.

A special case is the number 10, because its powers—$10^1 = 10$, $10^2 = 100$, $10^3 = 1{,}000$, $10^4 = 10{,}000$, and so on—are the basis of our numbering system, including decimal fractions, or *decimals* as they are usually called.

A decimal could be expressed like any other fraction, as a fraction whose denominator is some power of 10—e.g., $\frac{3}{10}$, $\frac{35}{100}$. But in our notation, the decimal is written as a number to the right of the decimal point. This number is the numerator, and the denominator is indicated by the number of places to the right of the decimal point. For example, 0.1 is the same as $1/10^1$, just as 0.01 is $1/10^2 = 1/100$; or 0.29 is $29/10^2 = 29/100$, and 0.0055 is $55/10^4 = 55/10,000$.

Expressed in another way, each digit to the right of the decimal point represents an ordinal number (or place value):

```
0 . 0 0 0 1
│   │ │ │ │
│   │ │ │ └ ten thousandths
│   │ │ └── thousandths
│   │ └──── hundredths
│   └────── tenths
└────────── units
```

Example:

$$0.1 = 1 \text{ tenth} = \frac{1}{10}$$

$$0.11 = 11 \text{ hundredths} = \frac{11}{100}$$

$$0.111 = 111 \text{ thousandths} = \frac{111}{1000}$$

$$0.1111 = 1,111 \text{ ten thousandths} = \frac{1,111}{10,000}$$

The addition of zeros after the last digit in a decimal does not change its value: thus $0.9 = 0.90 = 0.900$.

Important: A decimal is implied at the end of any whole number. Thus $100 = 100. = 100.0$

THE ELECTRONIC CALCULATOR

Everyone who deals with numbers is likely in our day to shift as much as possible of the burden of calculation to the ever-ready electronic calculator, now available in cheap and handy form, with its ability to present instant results of complicated operations, accurate to many decimal places—as many as seven places in the familiar eight-place calculator, or five places in the six-place calculator.

The calculator makes and delivers any fraction as a decimal. Thus the readiest way to carry out operations with fractions is to put them promptly into decimal form. Any fraction can be converted to an equivalent decimal by simply carrying out the division that it represents:

$$\frac{1}{4} = 1 \div 4 = 0.25 \text{ on the calculator}$$

$$\frac{3}{8} = 3 \div 8 = 0.375$$

$$\frac{5}{3} = 5 \div 3 = 1.6666666 \ldots; \text{ say } 1.67$$

$$\frac{38}{210} = 38 \div 210 = 0.1809532 \ldots : \text{ say } 0.18 \text{ or } 0.181$$

The endless fractions show up because some integers do not divide without remainder into 10 or any of its powers. They are easily dealt with. Either we take advantage of the calculator's "memory" to store the long decimals and carry on calculations until we reach a result where for other purposes we need to produce a usable number with only two or three decimal places. At that point we round off our result to the nearest hundredth or thousandth. Or we round off each continuing decimal that shows up as we go.

Rounding off. To shorten the seven- or five-place decimal that may appear in the calculator's "printout," we need to *round off*—i.e., drop the numerals at the end one by one until the result is of usable length—say three or two decimal places. Here is the simple rule. The last numeral is simply dropped if it is 0, 1, 2, 3, or 4. If it is 5, 6, 7, 8, or 9, it is also dropped, but the new last numeral is raised by 1. Thus 1.3333333 is reduced step by step—1.333333, 1.33333, 1.3333, to 1.333 or 1.33; and 1.6666666 goes from 1.666667 to 1.66667 and 1.6667 to 1.667 or 1.67. Likewise, 0.1809532 moves from 0.180953 to 0.18095 to 0.1810 and 0.181 and 0.18. The numeral 5 at the end is a rather special case, and dropping it can be determined by special rules, but we need not consider them here. For practical purposes, the above rule will apply.

Example: Consider the example under *Multiplying fractions,* above. With the calculator
$$\frac{1}{2} \times \frac{2}{3} \times \frac{5}{6}$$
becomes either 0.5 × 0.6666666 × 0.8333333 = 0.2777777, say 0.278; or 0.5 × 0.67 × 0.83 = 0.27805, say 0.278.

In the former case, we do not bother to round off the factors. We turn them over to the calculator and its memory, for which seven decimal places are as easy to work with as one, and we let it do the work. We may need to round off the result—even to 0.28 or 0.3. In the latter case, we round off to two places as we multiply and then round off the result as necessary.

Example: Consider again the example under *Division of fractions,* above.
$$\frac{1}{4} \div \frac{1}{5}$$
With the calculator, $\frac{1}{4}$ becomes 0.25 and $\frac{1}{5}$ becomes 0.2. Thus:
$$\frac{1}{4} \div \frac{1}{5} = 0.25 \div 0.2 = 1.25$$
There is no need to work with reciprocals.

PERCENTAGES

The term "percent" (%) signifies a specified number of parts of a whole divided into one hundred equal parts. For convenience and to prevent errors, mathematical manipulations require that a number occurring as a percentage—e.g., 5%— be expressed as a decimal, e.g., 0.05. A percentage is converted to its decimal equivalent by moving the decimal point two places to the left. Thus:

$$5\% = 0.05 \text{ (Remember: 5 is also 5.0)}$$
$$29\% = 0.29$$
$$150\% = 1.50$$

Moving the decimal point two places to the left accomplishes the same purpose as dividing the percentage number by 100, as

$$5\% = \frac{5}{100} = 0.05$$

Likewise, a decimal (such as 0.05) can be converted to a percentage by moving the decimal point two places to the right and adding the percentage sign, or multiplying by 100 and adding the percentage sign. Thus;

$$0.05 = 5\% \quad \text{or}$$
$$0.05 \times 100 = 5\%$$

EQUATIONS

An equation is a mathematical statement that certain things equal or balance each other. In radiologic technology our need is to understand basic equations which contain only one unknown. An equation is used to compare (balance) items and these items may be similar or dissimilar. An equation can be written to show that 12 eggs equal one dozen eggs:

$$12 \text{ eggs} = 1 \text{ dozen eggs}$$

In this case "eggs" is a constant term on both sides of the equation, but modified or described in a different manner—by "12" on one side, and "dozen" on the other. Since there are no other modifying terms, we assume that 12 must equal one dozen, which we all know is true. But suppose we do not know that 12 = one dozen, and let the term x represent the number of a "dozen"; then the equation can be written:

$$12 \text{ eggs} = x \text{ eggs}$$

The question then arises, What number does x stand for? To solve for x, knowns have to be placed on one side of the equation and unknowns on the other; thus,

$$12 \text{ eggs} = x \text{ eggs}$$

Dividing both sides of the equation by the constant term "eggs," we can cancel terms and obtain:

$$\frac{12 \cancel{\text{ eggs}}}{\cancel{\text{eggs}}} = \frac{x \cancel{\text{ eggs}}}{\cancel{\text{eggs}}}$$

or

$$12 = x$$

Therefore one dozen equals 12, or $x = 12$

This oversimple example brings forth some rules that must be followed in solving equations.

1. Each side of the equation may be added to, subtracted from, multiplied, squared, or treated with any other mathematical manipulation as long as the same thing is done to the other side of the equation at the same time.
2. A fractional equation $a/b = c/d$ may be cross-multiplied.
   For example,

$$\text{if } \frac{a}{b} = \frac{c}{d}, \quad \text{then } \frac{a}{b} \times \frac{c}{d} \quad \text{or } ad = cb$$

3. To solve equations, all knowns should be placed on one side of the equation, and unknowns on the other. By convention, the unknowns are usually placed on the left side of the equation, or are moved to that side. Moving the terms in an equation is accomplished by multiplying, adding, subtracting, or dividing both sides by the same number.

Example: $x - 4 = 10$. Find x.

Solution: To cancel the -4 on the left side of the equation, add $+4$ to each side of the equation. Thus:

$$x - 4 \ (+4) = 10 \ (+4)$$
$$x = 14$$

Proof: $14 \ (-4) = 10$.

In this case opposite signs in front of *like terms* cancel each other or reduce to zero. This has the effect of removing the (-4) from the left side of the equation and moving it to the right, leaving the unknown by itself on the left.

Example: Let $x + 4 = 10$. Find x.

Solution: Subtract 4 from each side:

$$x + 4 \ (-4) = 10 (-4)$$

Canceling and combining terms,

$$x = 6$$

Example: Let $4x = 10$. Find x.

Solution: Divide each side by $+4$:

$$\frac{4x}{4} = \frac{10}{4}$$

Cancel terms: $\dfrac{\cancel{4}x}{\cancel{4}} = \dfrac{10}{4}$

Thus, $x = \dfrac{10}{4}$ or $\dfrac{5}{2}$ or 2.50

Proof: $4(2.5) = 10$

Example: Let $x/4 = 16$. Find x.

Solution: Multiply each side by $+4$:

$$\frac{x}{4} \times \frac{4}{1} = \frac{16}{1} \times \frac{4}{1}$$

Cancel like terms: $\dfrac{x\,(\cancel{4})}{\cancel{4}} = 16(4)$

Thus, $x = 64$

Proof: $\dfrac{64}{4} = \dfrac{16}{1} = 16$

RATIO

"Ratio" is another term for "fraction" and serves the same purpose; it expresses the relationship between quantities in terms of numbers. The symbol used to indicate ratio is the colon (:) and is read "is to"; thus, the ratio 6 : 7 means 6 "is to" 7, or 6/7. The reverse is also true: any fraction can be written as a ratio: 6/7 = 6 : 7.

It is usual to express a ratio as a fraction and then carry out the desired mathematical manipulation on the fraction.

Example: Simplify the term 5/6 : 7/8.

Solution:

$$5/6 : 7/8 = \frac{\dfrac{5}{6}}{\dfrac{7}{8}} = \frac{5}{6} \times \frac{8}{7} = \frac{40}{42} = \frac{20}{21}$$

Remember: when dividing by a fraction, use the reciprocal of the divisor and multiply. Thus 5/6 : 7/8 is the same as 20 : 21.

Solution: In decimal form (as with the electronic calculator)

$$5/6 : 7/8 = 0.8333333 : 0.875 = 0.8333333 \div 0.875$$
$$= 0.9523809 \text{ or } 0.952$$

Note: $20/21 = 0.9523809$ or 0.952

Only the final result need be rounded off if we make efficient use of the calculator's memory.

Ratio problems can be solved by developing simple equations and then solving for an unknown.

Example: A man purchases three pieces of equipment for $30,000. The first piece cost $\frac{1}{4}$ more than the second, and the third cost $\frac{1}{2}$ more than the second. How much did each piece of equipment cost?

Solution: Step 1. Let the cost of the second machine be represented by x; then the first machine cost $1\frac{1}{4}x$ and the third cost $1\frac{1}{2}x$. All three machines cost a total of $30,000. An equation can then be set up in this form:

$1\frac{1}{4}$ x plus x plus $1\frac{1}{2}$ $x = $30,000$

Step 2. Simplify and add the terms, thus, transforming them into fractions with a common denominator:

$$\frac{5}{4}x + \frac{4}{4}x + \frac{6}{4}x = \$30,000$$

Simplify:

$$\frac{5+4+6}{4}x = \$30,000$$

Step 3. Multiply each side by 4 to get rid of the fraction.

$15x = \$30,000 \times 4$

$15x = \$120,000$

$x = \$120,000 \div 15 = \$8,000$

Thus, the cost of the second machine, x, is $8,000; that of the first machine ($1\frac{1}{4}x$) is $10,000; and that of the third machine ($1\frac{1}{2}x$) is $12,000.

Solution: $1\frac{1}{4}$ $x + x + 1\frac{1}{2}$ $x = \$30,000$. What is the decimal solution? Putting the terms in decimal form:

$1.25x + 1x + 1.5x = \$30,000$

$3.75x = \$30,000$

$x = \$30,000 \div 3.75 = \$8,000$

$1.25x = \$10,000 \quad \text{and} \quad 1.5x = \$12,000$

PROPORTIONS

A proportion is a statement that two ratios are equal. A proportion is indicated by the use of either the proportion symbol (::) or the equals sign (=). For example, a proportion is written as $a : b :: c : d$ and is read "a is to b as c is to d." As with ratios, this proportion may also be written in its fraction form. Thus,

$$a : b :: c : d \quad \text{becomes} \quad \frac{a}{b} = \frac{c}{d}$$

In the proportion $a : b :: c : d$, the outside terms, a and d, are called *extremes*, and b and c, the inside terms, are called the *means*.

There are a number of rules for operating with proportions which it is important to understand, since proportions are often encountered in radiography, particularly with the inverse square law which will figure prominently in Chapter 7.

Rule 1. In any proportion, the means can be exchanged without affecting the proportion. If $a : b :: c : d$, then the proportion can be changed to read $a : c :: b : d$ without changing the value of the equation.

Rule 2. In a proportion, the extremes may be interchanged without affecting the proportion. If $a : b :: c : d$, then $d : b :: c : a$.

Rule 3. The product of the means equals the product of the extremes. Again, if $a : b :: c : d$, then $b \times c = a \times d$.

The above rules can be proved by converting the proportion to its fractional equivalent. Thus $a : b :: c : d$ becomes $\dfrac{a}{b} = \dfrac{c}{d}$. As with any fraction, the expression can be cross-multiplied, and both sides added to, subtracted from, and divided without affecting the value of the equation. As noted above for proof:

Rule 1. Means can be exchanged.

Let $\dfrac{a}{b} = \dfrac{c}{d}$

Multiply each term by a constant $\left(\dfrac{b}{c}\right)$, then cancel terms, $\left(\dfrac{b}{c}\right)\dfrac{a}{b} = \left(\dfrac{b}{c}\right)\dfrac{c}{d}$

We obtain:

$\dfrac{a}{c} = \dfrac{b}{d}$ or $a : c = b : d$. The b and c terms have been exchanged.

Rule 2. Extremes can be interchanged.

Let $\dfrac{a}{b} = \dfrac{c}{d}$

Multiply each term by a constant $\left(\dfrac{d}{a}\right)$, thus canceling terms,

$\dfrac{d}{a}\left(\dfrac{a}{b}\right) = \left(\dfrac{d}{a}\right)\dfrac{c}{d}$

We obtain:

$\dfrac{d}{b} = \dfrac{c}{a}$, or $d : b :: c : a$

Rule 3. The product of the means equals the product of the extremes.

Let $a : b = c : d$, or

$$a/b = c/d$$

Multiply each term by bd. Cancel:

$$\dfrac{a}{b} \times bd = \dfrac{c}{d} \times bd$$

ad (extremes) $= cb$ (means)

THE METRIC SYSTEM

The United States is at last joining the rest of the world in using an international system of weights and measurements. This system, called the metric or decimal system, is based on the meter as the unit of length and the gram as the unit of weight. Volume is related to the liter, a liter being defined as the volume of a kilogram of distilled water at 4° C. To understand and to use the metric system, one has to think in decimals. To learn the decimal system requires one to commit only a few items to memory, but to use the decimal system requires us to discard the old British system of weights and measures (foot and pound). Hence, one must think in kilometers rather than miles, in centimeters rather than inches, and so on. Likewise, one must think in terms of decimals rather than fractions. (To convert a fraction into its decimal equivalent, one simply divides the numerator of the fraction by the denominator.)

The process of measurement assigns a number to some physical attribute of an object, such as length, volume, mass (weight), or time. Measurements are quantified descriptions.

It is easy to imagine how the "foot" was derived, but like other measurements, it varied from one person to another. In western Europe, it was the function of the local kings to set their country's standards of measure. Usually they chose typical body measurements—often the king's personal measurements: King Henry I of England (1068–1135) defined the yard as the distance from the tip of his nose to the end of his thumb when his arm was fully extended to one side!

At the same time, a system of mass measurements developed, using stones and crude balances. There was no effort to establish a relationship between units of length and mass. To avoid confusion, different units of measurement such as the furlong, rod, short ton and long ton gradually emerged and were standardized. But this haphazard development of a measurement system left unfulfilled the need for a well-ordered, interrelated network of measures.

Against the old tangle of unrelated units, the metric system is elegantly simple. Its units are related by powers of 10. A consistent system of prefixes designates multiples and subdivisions of the basic units. The units used in measuring different physical attributes are often interrelated.

The metric system came about in the course of the French Revolution, as the new French Republic undertook to replace the jumble of provincial standards with a single national system. In 1791 a committee headed by the mathematician Legrange and the great astronomer Laplace undertook to establish the new inter-related standards, all based on a unit of measure, the meter. (Originally the meter was carefully defined as a certain fraction of the earth's circumference. This basis turned out to be uncertainly measurable, but the chosen length, 39.37 inches, remained as the arbitrary standard.)

With the victories of Napoleon, the metric system spread through western Europe, replacing the countless local sets of unrelated units. The diffusion contin-ued even after hostilities ceased and Napoleon's empire fell apart. Commercial relations demanded uniformity of standards. When scientific forums began meet-ing again, communications were made difficult by standards that differed from scientist to scientist. Thus in science and commerce, the metric system was adopted in all European countries in the course of the nineteenth century—that is, all except the British Empire, which held fast to its Imperial system, with its inches, feet, yards, ounces, pounds, quarts, etc.—not to mention rods, furlongs, fathoms, drams, grains, short tons, long tons, pecks, bushels and so on.

Ever since the wave of enthusiasm early in the last century swept most nations into the fashion, the metric system has been used in science. The United States, newly founded, had its decimal impulses, too, when the time came to set up a monetary system. Our dollars, dimes, and cents are a model of regularity compared with other measures that the country took over from Great Britain and which it has held on to longer than any other major nation. At last in the 1960's Great Britain set about a changeover—beginning with the money system—to decimals and meters, with completion scheduled for the late 1970's.

METRIC CONVERSION FACTORS

The compelling reason at the heart of the metric system remains its system of interrelated units, all related to one another as powers of 10. Yet, because at the present time of transition we have to consider a mixture of British and metric units, it is essential to memorize some key conversions of the old units into metric units.

English		*Metric*
1 inch	2.54	centimeters
1 foot	30.5	centimeters
1 yard	91.4	centimeters or .914 meters
40 inches	101.6	centimeters or 1.016 meters
72 inches	182.88	centimeters or 1.83 meters
1 ounce	30	milliliters
1 pint	500	milliliters *
1 quart	1000	milliliters *

Note: In radiography a 40-inch focal film distance is equal to approximately 1 meter; 72 inches is approximately 1.8 meters.

In the metric system cubic centimeters (cc) are used for gas volume, whereas milliliters (ml) are used for liquid volume. Thus a patient is given x milliliters of a radiographic contrast media, but x cubic centimeters of carbon dioxide is used for a pararenal insufflation study. For practical purposes 1 milliliter equals 1 cubic centimeter.

Although not a part of the metric system, the term "hertz" is used to designate 1 cycle per second. So for the 60 cycle per second power supply customary in the United States, we now say 60 hertz (abbreviated Hz) instead of 60 cycles

* Actually 1 pint equals 473 ml and 1 quart equals 946 ml. However, for practical purposes, a pint equals ½ liter (500 ml) and a quart equals 1 liter (1000 ml).

per second. Along with this, the x-ray pulses are described in the decimal rather than the fractional system. Thus, in a 60-Hz power supply,

    1 cycle per second $=$ 1 hertz
    ½ cycle per second $=$ 0.5 hertz

For timing purposes;

        2 pulses $=$ 1 hertz, which requires 16.67 milliseconds (⅟₆₀ second)
        1 pulse $=$ 0.5 hertz which requires 8.33 milliseconds (⅟₁₂₀ second)

Centigrade temperatures

Related to the metric or decimal system are temperature readings. Anders Celsius, finding the Fahrenheit scale difficult to work with, arbitrarily assigned 0° to the freezing point of water and 100° to the boiling point. He then subdivided the temperature differences between boiling and freezing into 100 degrees. This system, called the Celsius system, is also known as the centigrade (C) system because there are 100 degrees (centi $=$ 100) between the freezing and boiling points of water. In radiologic technology we are concerned with temperatures ranging from room temperature to 100°C. The temperatures frequently encountered are listed below.

Water freezes	0°C
Chilly day	5°C
Comfortable room	22°C
Hot summer day	33°C
Normal automatic processing	33–34°C
Normal body temperature	37°C
Water boils	100°C

To convert from Fahrenheit to centigrade temperatures, this formula is commonly taught:

$$X°F = \frac{5}{9}(X - 32)°C$$

Example: An automatic film processor is operating at 92° F. What temperature is this in the centigrade system?
Solution:

$$92°F = \frac{5}{9}(92 - 32)°C$$
$$= \frac{5}{9} \times (60)°C$$
$$= \frac{5\,(60)}{9}°C$$
$$= \frac{300}{9}°C$$

Therefore, 92°F $=$ 33°C

Conversion from the Fahrenheit to the centigrade scale is easier if the ratio concept is applied. For example, we know that water boils at 212° F (100° C) and freezes at 32° F (0° C). By forming a relationship between boiling and freezing points for both the Fahrenheit and centigrade scales, we see that

$$\frac{212°F \text{ (boiling point)} - 32°F \text{ (freezing point)}}{100°C \text{ (boiling point)} - 0°C \text{ (freezing point)}} = \frac{180}{100} = 1.8$$

Thus,

$$°F - 32 = °C \times 1.8 \quad \text{or}$$
$$°F = (°C \times 1.8) + 32 \quad \text{or}$$
$$°C = \frac{°F - 32}{1.8}$$

In this textbook the metric system will be used as far as possible. The student should be thoroughly familiar with the system before proceeding further.

PROBLEMS IN SIMPLE MATHEMATICS

1. Convert the following fractions into their decimal equivalents:

   a. $\frac{1}{2}$

   b. $\frac{1}{30}$

   c. $\frac{1}{60}$

   d. $\frac{1}{120}$

   e. $\frac{1}{10}$

2. In a 60-cycle, single-phase x-ray generator,
   a. 1 pulse = ____ ms;
   b. 2 pulses = ____ ms.

3. In a 60-cycle, 3-phase, 6-pulse generator,
   a. 1 pulse = ____ ms;
   b. 2 pulses = ____ ms.

4. In a 60-cycle, 3-phase, 12-pulse generator,
   a. 1 pulse = ____ ms;
   b. 2 pulses = ____ ms.

5. Convert the following from the English to the metric system. Show your calculations.
   a. 22 in. = ____ cm;
   b. 72 in. = ____ cm;

    c. 36 in. = _____ m;

    d. 5 lbs. = _____ kg;

    e. 6 fluid ounces = _____ ml;

    f. 95°F = _____ °C;

    g. 68°F = _____ °C.

6. Rule: The milliamperage (mA) required for a given exposure is inversely proportional to the time *(t)*.

$$\frac{mA_1}{mA_2} = \frac{t_2}{t_1}$$

In the chart below, fill in the missing items.

mA_1	mA_2	t_1	t_2
30	___	500 ms	50 ms
50	25	100 ms	___

7. Rule: The intensity or exposure rate of radiation at any given point is inversely proportional to the square of the distance from the source. Hence,

$$\frac{mA_1}{mA_2} = \left(\frac{d_2}{d_1}\right)^2$$

In the chart below, fill in the missing items (SID = source/image-receptor distance).

mA_1	mA_2	SID_1	SID_2
100	___	100 cm	50 cm
100	200	100 cm	___

8. A urogram was being performed on a large patient using a 300-mA single-phase generator. An exposure of 5 mR (milliroentgens) produced an underexposed radiograph. The patient was too agitated for the technologist to be able to increase the time of exposure. Kilovoltage could not be increased because of the need for contrast in order to visualize the contrast media. The technologist decided to decrease the distance between the x-ray tube and the film from 100 cm to 80 cm. At this new distance, what would be the exposure of the patient to radiation?

9. A phototimer was set on the 300-mA station where exposures were noted to be extraordinarily long. The technologist changed the phototimer station to the 500-mA station. What would happen to the time of exposure?

2 RADIOGRAPHIC IMAGE RECEPTORS

Radiographic film is the principal medium used by radiologic technologists to record information that will be used by others in forming opinions or diagnoses concerning the presence or absence of disease processes. Film is not the only medium used to record medical information; electronic recording is possible with the aid of an image intensifier. Now that electronic imaging is evolving, other and even greater developments may be expected. In the future, changes in radiographic film itself may take place, particularly in size and format. Since radiographic film receives an image, as does an image intensifier, we can speak of both as *image receptors*.

Radiography is a specialized field of photography. Both use radiant (electromagnetic) energy to record an unseen or latent image which can be transformed later into a visible, useful image. In photography, light rays reflected from an object are gathered by a set of lenses and focused upon a recording medium, the film; in radiography, energy passed through an object is used to produce an image. By using lenses and filters the image can be manipulated in photography; in radiography, the x-ray beam cannot be altered except for size, quality, and quantity. Photographic film has an emulsion only on one side of its base; in radiography, both sides of the film are generally coated with an emulsion. In photography, long-wavelength energies in the visible light spectrum are used; in radiography, short wavelengths in the x-ray or gamma spectrum are used. Light produced by solar bodies for photography is abundantly available. In contrast, the energies required to produce x-rays are grossly inefficient; less than 1 percent of the energy supplied to an x-ray tube is made available to assist in forming an image. However, the process in which radiographic film is exposed is fundamentally the same as the photographic process.

The end result of photography, the photograph, is produced through a negative-positive process. The radiograph is a negative; that is, the light and dark tones or shades in the image are the reverse of the tones in the original subject; the blacks in the image correspond to whites in the original, and the whites to blacks. In a photographic negative, this reversal is the result of exposing the film by reflected light. Dark parts in the object being photographed absorb more light than bright parts; thus, dark parts reflect less light to the camera. In radiography, a radiograph records energy which passes through an object. The intensity of the energy decreases proportionately as the various densities in the object absorb different amounts of energy when the beam passes through. Generally speaking, a radiograph is always a negative; exceptions are Polaroid and duplication film,

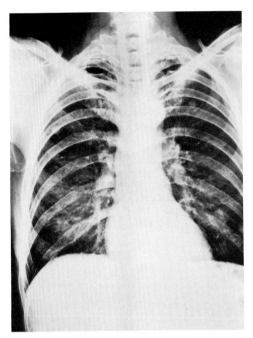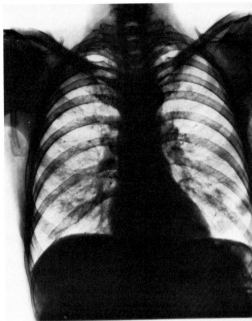

Figure 2-1. Comparison between a negative and a positive film. On the left is a negative film. All of the blacks and whites are reversed in tone as compared with the positive film on the right. A number of European publications use positive rather than negative prints.

which produce a positive radiograph or copies, respectively.* To observe the difference, refer to Figure 2-1. In the radiographs produced as illustrations in this book, notice how the opaque parts of the body—particularly the bone—show up as lighter-shaded areas. This is due to the higher density of some tissue absorbing a proportionately larger amount of the x-ray energy. On the other hand, dark areas on the film are due to a proportionately larger amount of x-ray energy passing *through* the tissue with very little absorption. Soft tissue densities such as muscles, nerves, and fluid produce a series of gray tones in the film since they absorb less than bone but more than air. Bone appears as a low density (white) on the radiograph whereas air corresponds to a high density (black).

THE STRUCTURE OF RADIOGRAPHIC FILM

Radiographic film is composed of a base, an emulsion, and a protective coating (Figure 2-2).

* Duplication film is a positive copy of a negative film. Therefore, when the film is processed, it is also a negative.

18

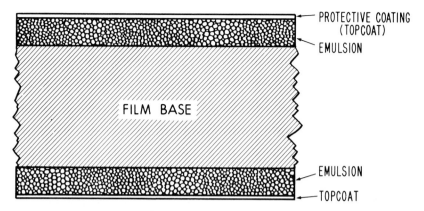

Figure 2-2. Cross-sectional drawing of a radiographic film. The labels are self-explanatory.

The film base

The base is composed of a polyester which measures 6.8 to 7 mils thick (a mil is a thousandth of an inch). Except for ciné film, cellulose acetate is no longer used in manufacturing radiographic film. Polyester has the advantage over the old acetate base that it shrinks less when put in water, holds less moisture, and can be made thinner than acetate bases while maintaining the rigidity necessary for easy handling. By making the base thinner, the emulsions, coated on both sides of the base, are closer together, thus decreasing "crosstalk."

Not all of the light absorbed by the film emulsion is generated by the screen adjacent to the emulsion; some of the exposure comes from light photons generated in the opposite screen. This exposure caused by photons from the opposite screen is called *crossover* or *crosstalk*. As the light photons penetrate and spread through the film base to the opposite side, the image also spreads, and thus the image sharpness is degraded. Crosstalk can be controlled somewhat by making the base thinner and by adding special dyes to the base which absorb a portion of the unwanted light photons. Dyeing the base also improves visual perceptibility.

The emulsion

The emulsion is composed of a binder and a recording medium. The binder holds the recording medium on the base and provides a form of material easier to work with. The binder is gelatin, which is extracted from collagen, a protein obtained from the flat bones of cows. Gelatin is classified chemically as a colloid. A colloid is a substance which, in solution, fails to settle out but imbibes (absorbs) water. At normal processing temperatures, a colloid will not dissolve; at higher temperatures it will. A colloid also has some degree of porosity. The emulsion (gelatin) must swell in water so as to allow the processing chemicals to penetrate the emulsion and reduce the exposed silver halide crystals to black silver metal (in the developer) and remove the unexposed silver halide crystals (in the fixer) during processing. As the film dries, the binder also has to release water and return to its original thickness. Processing also stabilizes and hardens the emulsion, thus allowing the film to be kept a long time without change in quality.

The gelatin used as a binder in radiographic film is similar to the Jello used in home cooking, though much more highly purified. When the powder is put in hot water it dissolves, and when the solution cools, it jells, that is, takes a form. If Jello is allowed to stand, it will absorb water from the air and liquefy. As it dries, it assumes a more rigid form, just as gelatin does in radiographic film.

The halide salts of silver are the usual recording media in photography, and in radiography a silver bromide emulsion is the most commonly used. Other silver salts used in the imaging process include the iodide and the chloride. Silver bromide is used because it is most sensitive to the color of light emitted by intensifying screens. Silver bromide is made by dissolving metallic silver in nitric acid to form silver nitrate. The silver nitrate combines with potassium bromide to form a silver bromide precipitate. This precipitate is mixed with the gelatin binder to form the film emulsion.

Film grain

During the manufacturing process, silver halide particles are dispersed throughout the gelatin. An attempt is made by the manufacturer to disperse these particles evenly. However, silver halide crystals tend to clump, and this clumping produces what is known as "film grain." During manufacture, film speed is determined and controlled by adding the right amount of different dyes to the film emulsion. The more dye there is added to the emulsion, the more light photons there will be absorbed by the film emulsion, and thus, the slower the film; and vice versa. Film speed is further controlled by making the silver halide crystals more sensitive or less sensitive to a light photon and by controlling the number of silver atoms produced by that light photon. A single photon will produce much less film blackening on a slow film than on a fast film. Slow films tend to have much less clumping of silver halide crystals in the emulsion than faster film emulsions. Thus, fast films tend to be much more grainy than standard-speed film. Direct-exposure mammographic film, on the other hand, tends to have very little graininess and is known as fine-grain film.

The protective coating

After the emulsion is coated on the base and allowed to dry, a thin protective coat is applied on top of the emulsion which serves to protect it from scratches and other harmful agents, as well as to prevent glare when the finished radiograph is viewed.

EXPOSURE AND LATENT IMAGE FORMATION

It is outside of the scope of this text to discuss exposure and latent image formation in detail, but we review these topics here briefly. For more specific detail, please refer to standard reference books on the subject listed at the end of this chapter.

After the silver bromide solution is mixed with gelatin to form the emulsion which is coated on a polyester base, this emulsion must be exposed to some type of energy before an image can be produced. Many types of energy can expose radiographic film—chemicals, heat, and radiant energies such as light, x-ray, or gamma rays. Our interest is primarily with x-ray exposure.

In radiography, the production of an image begins when x-ray photons or light photons produced by an intensifying screen strike a silver bromide crystal contained within the film emulsion. Silver bromide crystals have an abundance of bromide ions. When these ions are struck by an appropriate level of energy they emit electrons which migrate to a minute particle in the silver halide crystal. These particles are called *sensitivity specks,* or centers. There the electrons (each a negative electrical charge) attract positive silver ions, which are deposited as metallic silver atoms. Activation of the sensitivity speck allows the developer later to reduce all the silver atoms in the crystal, thus increasing the number of reduced silver atoms by over one million times. This movement of bromide ions to the sensitivity speck is also known as the *entrapment stage* (in the Gurney-Mott theory), and the movement of the silver ions and their deposition is known as the *migration phase.* This ionization and deposition of silver produces what is known as a *latent image.* The invisible latent image is made visible by the process of developing the film.

On being exposed, the film becomes sensitized and is much more sensitive to further exposure than before. It remains so until processing stabilizes it. While the period of sensitization lasts, the film is much more prone to be excited by low levels of radiation of any sort, including light from safety lamps in the darkroom. The amount of sensitization will depend on numerous factors—temperature of the film, type of exposing energy, type of film, and so on.

DEVELOPMENT OF THE VISIBLE IMAGE

Film is developed at an appropriate time-temperature-chemical activity relationship which is discussed in depth in Chapter 11. The higher the temperature, the less the time required to develop the film. The more active the chemicals, the less the time or the lower the temperature required to properly develop the film. The chemicals used in developing the film, collectively called the developer, are composed primarily of mixtures of phenidone and hydroquinone. Phenidone acts quickly, and its purpose is to develop low contrast levels in the film. It is very sensitive to temperature changes and primarily controls film speed and low density contrast. Hydroquinone acts much more slowly and is not as sensitive to time-temperature changes. It is used to develop the maximum density of the film and film contrast. The two chemicals act synergistically, i.e., the resultant reaction is greater than what would be expected if one simply added the two reactions together.

Other chemicals are added to the developer. Potassium bromide is used as a restrainer or antifogging agent; sodium carbonate is an activator which provides the alkaline medium required for the reducing agents and assists in swelling the gelatin in the emulsion. Sodium sulfite acts as a preservative protecting the developer from aerial oxidation. Glutaraldehyde is added to control the swelling and hardening of the emulsion. All of these chemicals are added to water, which provides the vehicle for a solvent base and aids in gelatin swelling.

The function of the developer is to selectively reduce the sensitized silver halide crystal. Developer attacks only sensitized silver halide crystals under controlled conditions. The developer provides electrons for the reduction of the silver halide crystal: by giving up electrons, the developer is oxidized; by gaining electrons, the silver halide is reduced. Aggregates of these reduced crystals form the visible image. Nonsensitized silver halide crystals are not affected by the developer and are removed from the emulsion by the fixer solution.

TYPES OF X-RAY FILMS

There are two basic types of x-ray films used in medical radiography—(1) regular-type film, used with or without intensifying screens, and (2) direct exposure film. The emulsions of each type are made to be most sensitive to a different spectrum of energy. Thus, screen-type film is faster when used with intensifying screens and slower when exposed by direct radiation. The opposite holds true for non-screen film. When regular-type film is used without intensifying screens (with cardboard holders), it is slower in speed than direct or non-screen film. Non-screen film is more sensitive to direct x-ray exposure and should not be used with screens. Non-screen film or regular film used in a cardboard holder is usually employed for small parts only. Because of its thick emulsion, non-screen film generally cannot be used with automatic processing and is therefore not used as much now as it was in the past. Regular-type film is more sensitive to the fluorescent output of intensifying screens than non-screen film. If one uses regular-type film in cardboard holders, one should bear in mind that the film contrast will be decreased as compared with using regular-type screen film with intensifying screens. Hence, lowering of contrast or increase in latitude allows one to see soft tissues better with direct exposures than with screen-type exposures.

Blue- and green-sensitive film

With the advent of rare earth high-speed intensifying screens, the terms panchromatic and orthochromatic have become familiar. By definition, panchromatic film is responsive to all colors in the light spectrum. Orthochromatic film is responsive to all colors except red. Standard x-ray films are blue-light sensitive. When this sensitivity is extended to include green, the film is orthochromatic. When the sensitivity is extended further, to include red also, then it becomes panchromatic. Medical x-ray film in the past was manufactured to be most sensitive to the spectral wavelength of photons emitted from calcium tungstate screens. This spectral emission is at a low level, in the region centered at about 420 nanometers (4200 angstroms), which is primarily in the blue spectrum. Light from some of the rare earth screens, particularly gadolinium oxysulfide and lanthanum oxysulfide, is in the green spectrum, primarily at 545 nanometers.

As will be discussed in Chapter 3, on intensifying screens and cassettes, some of the rare earth screens require the use of green-light sensitive film, and this film is usually called orthochromatic film. A standard blue-sensitive film is used with calcium tungstate screens. The trade name Ortho-G is often misunderstood, many thinking that Ortho-G film is sensitive only to the green-light spectrum and not to the blue. It is sensitive to both, but more so to green than blue, like any green-sensitive film. Ortho-G film, being more sensitive to the green-light spectrum of rare earth screens is also sensitive to the light emitted from a standard Wratten 6B darkroom safety light. Therefore, the darkroom safety light may have to be changed in order to prevent fogging of the film. The new GBX filter is reported to be usable for both Ortho-G and standard blue-sensitive films.

The whole subject of blue and green sensitivity and blue and green emission has been confused by some recent advertisements asking "Why green?" The student should know that screens that emit a green light also emit some blue light, but at a lower intensity. Screens that emit blue do not emit a significant amount of green.

Films are much the same. Green-sensitive film is nothing more than blue-sensitive film that has additional green sensitization to take advantage of the intense green emission peaks of some of the rare earth screens. Therefore, if a blue-sensitive film is put in a "green" rare earth screen system, it will still work, but at a considerably slower speed. The converse is also true: if a "green"-sensitive film is put in a blue rare earth screen system, it also will work, but at a slower speed, since its green sensitivity is not being utilized.

The result of selecting a film that is not sensitive to the spectrum of the intensifying screens being used is to decrease the speed of the film/screen system. Hence, one must use the film designed for a film/screen system; otherwise the patient will be exposed unnecessarily to x-radiation.

Mammographic film *

Mammographic film is a special film designed for radiography of the breast. It is a direct-exposure film capable of being developed in mechanical processors. This film can also be used in radiography of small extremities where detail is necessary.

Up until a few years ago, the standard radiographic film used in mammography was high-speed industrial film. Recently 90-second processible film has been introduced for mammography. The systems include both calcium tungstate and rare earth film/screen combinations. These systems require less exposure than industrial films, thus making possible a decreased exposure of the patient. The systems can be used with cassettes having special low-absorption fronts or in vacuum bags. Cardboard backing is added to provide some rigidity to the film/screen package. By using a vacuum bagger, a superb film/screen contact is possible. These new systems, in addition to being many times faster than high-speed industrial film, also have a better low-density contrast level. An x-ray exposure reduction is possible because the single screen/single emulsion absorbs and converts x-ray energy much more efficiently than any double silver halide emulsion alone.

Therapy localization film

Therapy localization film is a direct-exposure fine-grain film which gives acceptable radiographic detail under a wide range of exposures to x- or gamma rays. Depending on the type of film, it can be processed either manually or in 90-second processors.

Dental x-ray films

Dental x-ray films are available in a number of sizes, types, and speeds. The sizes may be classified as periapical, interproximal, occlusal, and panoramic.

Periapical dental film is packed in light-tight, moisture-proof packets that can be placed inside the mouth. It is useful in examining the roots of the teeth. The emulsions are usually classified as fast in sensitivity or speed. Interproximal or bite-wing film packets are used for localizing cavities between the teeth. Occlusal film is larger than the periapical film and is employed for examining larger dental areas that cannot be covered with periapical film. Panoramic film, which is nothing more than standard x-ray film cut to a size to fit a panoramic screen cassette, is used for x-raying the entire mouth.

---

* Mammographic film is not an extra fine grain film in a strict technical sense. Photographic scientists consider the grain to be intermediate in size. The perception of fine grain results more directly from the low level of quantum mottle, rather than from the grain size of the film. The film grain itself is usually perceptible only when the image is magnified by projection.

Personal monitoring film or dosimeter films

These films are designed for recording x-, gamma, or beta radiation exposure to personnel. Dosimeter film packets may contain one or more pieces of film for recording different types of radiation. Normally, a double-coated film shows exceedingly high sensitivity and another piece of single-coated film shows low sensitivity. The combination of film allows the recording of radiation exposure of the range of less than 13 mR (milliroentgens) to approximately 1800 roentgens of x- or gamma rays. A number of films are also available for measuring exposure to slow and fast neutrons. Dosimeter or monitoring films are usually in the form of small dental-size packets.

Industrial film

Industrial film can be used in mammography for direct exposure and can also be used in radiographing small parts of the body. Because exposure to radiation when using industrial-type film far exceeds that from a screen exposure, industrial-film applications are limited in medical radiography.

Ciné film

Ciné fluorography is the process of recording via the output phosphor of an image intensifier. Film used in this process is commonly known as ciné film. Ciné film comes in two sizes; 16 and 35 mm (millimeters). The millimeter size refers to the width of the film. Ciné film is usually green-light sensitive, since the output of an image intensifier is in the green-light spectrum. The film is a single-emulsion film and is perforated, that is, has sprocket holes. It may be high-speed or medium-speed. "Reversal film" produces a positive rather than a negative. The speed, contrast, and fog of the ciné film can vary greatly, depending on processing conditions.

Spot-film camera film

Spot-film cameras, also known as fluorospot or photofluorographic cameras, use film with widths of 70, 90, 100 or 105 mm. The 70, 90 and 105 mm widths are packaged in rolls, like ciné film. The 100-mm spot film is precut and is packaged in sheets. Fluorospot camera film also has a single emulsion.

Automatic serial changer film

As radiography becomes more and more automated, more changer film will be used. When speaking of automated equipment, one usually thinks of an automatic chest film unit or of an automatic radiographic table. In addition, one uses film changer film in such devices as the Elema Schonander and Franklin film changers. Film for serial film changers has generally the same emulsion as regular radiographic film, but in addition a special protective coating (top coat) is applied to the film to make it move more smoothly through automatic film devices and to help prevent the build-up of static. For all practical purposes, one generally has to use changer film in automatic devices, since nonchanger film tends to jam. Screen surface texture is more important than film top coat when transporting through mechanical devices. Film top coat is important to eliminate stacking errors in receiving magazines.

DUPLICATION AND SUBTRACTION OF RADIOGRAPHIC FILM

Duplicating radiographs

Duplicating radiographs is a task commonly assigned to a radiologic technologist within the x-ray department. There are a number of reasons why radiographs are duplicated, the more common being to make teaching aids or to supply radiographs to other institutions without lending the originals. Duplicating radiographic films is expensive, and a lot of money can be wasted if one is not thoroughly familiar with the process. It is obviously much cheaper to do it right the first time rather than having to make a number of exposures before producing a duplicate acceptably close to the original. Excellent copies of radiographs can be made with a variety of mechanisms, as long as one knows the exact procedure.

Duplicating film is a single-emulsion film, and the emulsion side must be placed against the original radiograph to maintain sharpness. It is a positive and can be processed either manually or mechanically. The film can be exposed in a printing frame like a commercial contact printer or in a box printer commercially available from many sources.

An ultraviolet light is the energy source for exposing duplicating film; generally this is a BLB ultraviolet fluorescent lamp, obtainable at most electrical supply houses. A standard light bulb can be used, but this has the effect of enhancing contrast. Since a positive film is being used in the duplicating process, one should remember that as exposure time is increased, darks become lighter. To increase blackness, exposure time must be reduced.

To make copies with a Blu-Ray copier, one need only feed the radiographic duplicating film into the printer. The film is automatically returned in about 6 seconds and is ready to be processed. Figure 2-3 is a photograph of the Blu-Ray machine. Figure 2-4 shows original radiographs and the duplicates made with a duplicating machine.

The Du Pont Cronex printer was designed to be used both for duplicating radiographs and for subtraction techniques. Since the printer is a direct-viewing apparatus, registration and duplication of the film become easier. The printer has three light sources available. One is for viewing to check registration, another is a white-light source used in making subtraction masks, and the third is an ultraviolet source for use in making duplicate radiographs. The device also has an automatic exposure timer, which allows exposures up to 15 seconds in one-quarter-second intervals. Figure 2-5 illustrates the Cronex printer.

Another method of duplicating radiographic film is with a LogEtronics printer or similar equipment (Figure 2-6). Instead of using full-size film, the radiograph is optically reduced so it will fit a 70, 90, or 105-mm format. The obvious advantage of reducing radiographs is that less film and chemistry are used in making the duplicates. Since the equipment costs more than the equipment required for a standard full-size duplicate, the duplicating process is more expensive. However, the small-format film is easier to handle and cheaper to mail than larger film. There are also obvious cost savings in the cost of the film and chemical supplies needed to process the film. But one needs some experience if one is to use small-format films successfully, both in handling and in viewing.

Subtraction techniques

The use of subtraction techniques is almost entirely limited to angiographic procedures. The purpose of subtraction is to remove, cancel out, or subtract unwanted information so that contrast-filled vessels are more clearly seen. The items usually subtracted are images of bony structures which overlie an area of interest. The subtraction technique requires two identical radiographs, one with and one without contrast media. Normally, the radiograph without contrast media is obtained as the first exposure from a series of film exposures made over a period of a few seconds. The film with contrast media is obtained from an exposure made further on in the exposure series. By using films obtained in this manner, there generally is no problem in getting identical position and exposure.

Figure 2-3. The Blu-Ray radiograph duplicating machine.

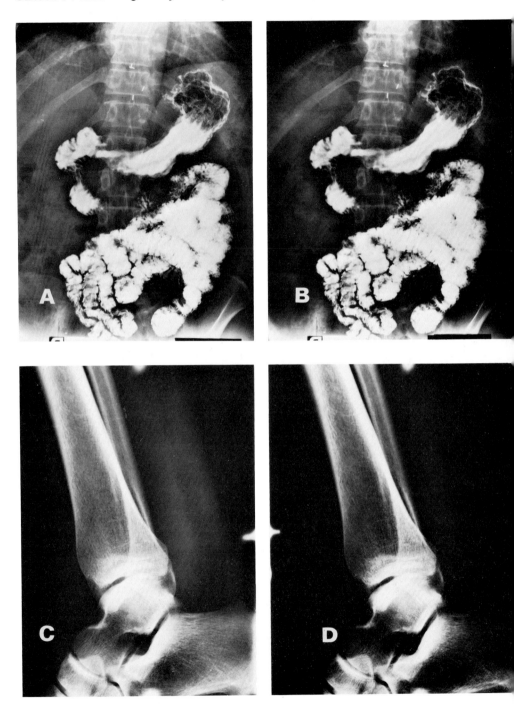

Figure 2-4. Two pairs of radiographs—originals and duplicates. Note that the originals, **A** and **C,** have slightly less density than their copies, **B** and **D.**

Cancellation (subtraction) of unwanted information is achieved with a subtraction mask. A subtraction mask is a contact duplicate which has a reversed tone image as compared to the original radiograph. A direct-viewing copier is ideal for subtraction, since proper registration is critical. Care must be taken that the reverse tones are as opposite as possible to the original.

To obtain the subtracted film, the subtraction mask is placed over the original radiograph in an exact register, i.e., the subtraction mask is aligned to the original radiographic counterparts as closely as possible. Poor registration causes a prominent white or black boundary edge of the image. By duplicating the subtraction mask and the original radiograph, the effect is to cancel out unwanted information. Figure 2-7 shows a subtracted film.

Subtraction film is a single-emulsion film. Since subtraction appears to work better with a slightly unsharp image, it is sometimes better to place the subtraction

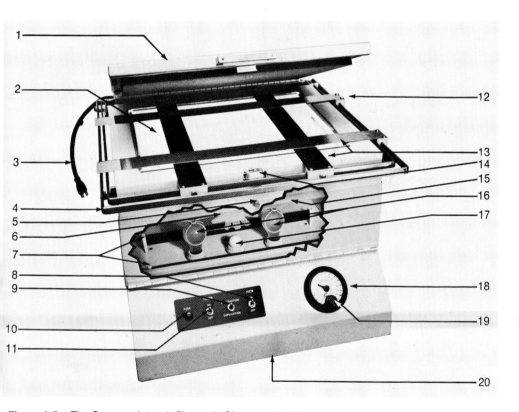

Figure 2-5. The Cronex printer: **1.** Platen. **2.** Glass top. **3.** 110-volt, 3-conductor power cord, **4.** Guide-bar frame. **5.** Lock screw (remove to open front panel). **6.** Intensity equalizer. **7.** Duplicating light source—2 black-light-blue tubes. **8.** View-light switch. **9.** Selector switch for subtraction or duplication. **10.** Fuse (4 amp). **11.** Timer on-off switch. **12.** Clamp block with thumb screw. **13.** Hold-down strips. **14.** Catch. **15.** Diffuser glass. **16.** View lights—two 75-watt bulbs. **17.** Subtraction light source—7½ watt bulb. **18.** Automatic timer. **19.** Timer start button. **20.** Ballast transformer (not visible but accessible from underneath unit). (Courtesy, E. I. du Pont de Nemours & Company, Wilmington, Delaware.)

Figure 2-6. LogEtronic model R-46 reducer. This piece of equipment is the heart of the LogEscan System and produces a minified 105 mm. copy of the original radiograph. Either a positive or negative can be made with the equipment, depending on the type of film used in the camera. Many of the radiographs in this book were reproduced with the R-46. (Courtesy, LogEtronics, Springfield, Virginia.)

film over the original radiograph with the emulsion side up rather than down as with duplicating film.

FILM STORAGE AND HANDLING

Film is a delicate material and should not be handled carelessly or roughly. It is sensitive to heat, light, x-rays, gamma rays, chemical fumes, pressure, rolling, bending, etc., and any one of these is capable of adversely affecting the emulsion. However, radiographic films can be handled safely by the x-ray technologist as long as one knows what must be done to avoid the production of foreign marks, referred to as *artifacts.* In radiographic terminology an artifact is a mark which is foreign to the x-ray image and is not necessarily imposed on the film by the action of x-rays.

Radiographic film is packaged in a photo-inert polyethylene bag or in metal foil to protect it from light and moisture. Film is packaged in different quantities up to 125 sheets per box. The volume of film used should be considered when choosing a package size, since humidity control is lost when the protective bag

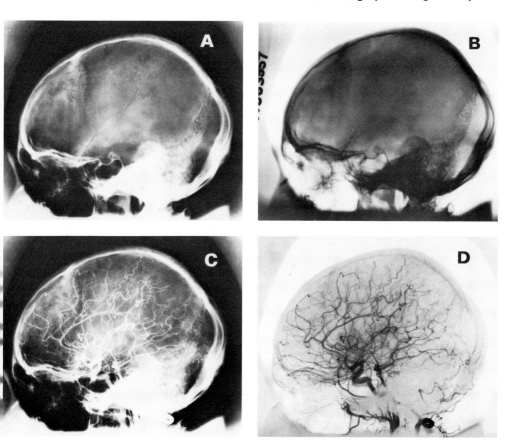

Figure 2-7. Subtraction technique. A subtraction mask **(B)** is made of the original base radiograph **(A)**. The subtraction mask is superimposed over the original angiogram **(C)** and the resulting print of the composite **(D)** clearly shows the angiographic vessels.

is opened. If the humidity is well controlled in the darkroom, package size becomes less important. However, if there are large changes in humidity while the package is being used, static electricity may develop and produce film artifacts. In cases of high humidity, films may stick together. Film sheets may or may not be separated by photo-inert leaves of paper; hence, the terms interleaved and non-interleaved.

Heat

Unexposed and unprocessed film should always be kept in a cool, dry place. It should never be stored in damp places, such as basements or near steam pipes, radiators, or other sources of heat. High temperatures increase the sensitization of the film emulsion, causing loss of contrast and the production of fog upon processing. Heat, in effect, causes a low-level exposure of the film.

Ideally, film should be kept in a cool, dry room with temperature ranges of

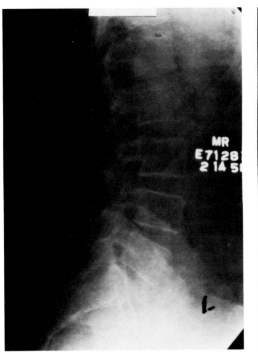
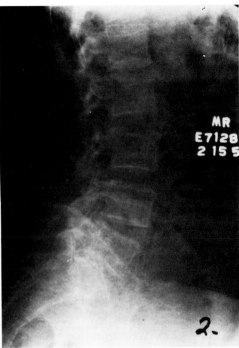
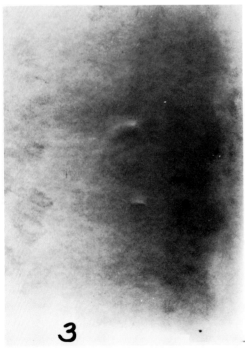

Figure 2-8. **(1).** An "overexposed" radiograph of the lateral lumbar spine. **(2).** A repeat film with 50 percent reduction in exposure factors: the film still appears overexposed. **(3).** Non-exposed film from the same box as **(1)** and **(2)**, showing that most of the density in **(1)** and **(2)** was contributed by fogged film.

10°C to 21°C and 40 to 60 percent relative humidity. The cooler the room, the longer the film may be stored without an objectionable increase in fog. Film cartons should be stored on edge to prevent development of pressure artifacts.

Radiation

Radiographic film must be suitably protected from the action of x-rays and all other forms of radiation. This can be accomplished by storing film in a lead-protected area or by keeping the film at a safe distance from gamma and x-ray exposure.

Expiration date

Film should be used before the expiration date, since film aging causes a loss in speed and contrast and increases fog. The expiration date will be found printed on the top or side of each box of film. The film with the nearest expiration date should be used first, newest film last. One should adhere to the first in/first out (FIFO) system; the oldest film is used first and the newest last.

Figure 2-8(1) shows a lateral lumbar spine thought to be overexposed. Figure 2-8(2) is a radiograph of the same patient with a reduction of 50 percent in exposure employing the same cassette. There is very little difference in the two films, since detail is obscured by a high film fog typical of outdated film. Figure 2-8(3) is a film taken from the same box of film and processed in the normal manner (without radiographic exposure). The high fog level is apparent. Unexposed film is normally clear.

A mottled appearance can be caused by a "white light" leak in the film bin. Interleaving paper when struck by white light gives mottled appearance to the film (Figure 2-9). Most film is now supplied without interleaving paper.

Handling

When handled, radiographic films should be held as near the edges as possible. The hands should be clean and dry, and hand cream should be avoided. The use of rubber gloves should also be avoided, since they tend to produce static electricity. *Persons whose hands tend to be moist should use cotton gloves when handling film.*

Types of film holders

Generally there are two types of containers in which medical x-ray film can be held during exposure—*cassettes* and *cardboard holders.* More recent is the disposable cardboard or paper holder. The choice of holder depends upon the part to be radiographed and the exposure technique. Loading and unloading film is of great importance and requires care. Film should never be slapped

Figure 2-9. Mottle caused by white-light exposure through interleaving paper.

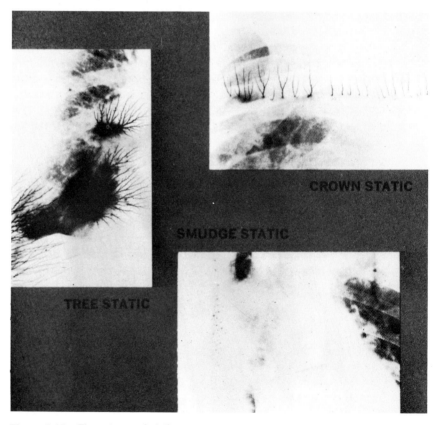

CROWN STATIC

SMUDGE STATIC

TREE STATIC

Figure 2-10. Three types of static.

into a cassette or jerked from it; and the cassette should not be slammed shut, since this may produce pressure artifacts.

Static electricity

When two nonconductors are separated quickly, a static charge may be established. Electrical charges may be generated when a cassette is opened and the two screen surfaces are rapidly separated, or when a film is pulled from the cassette. The rubbing together of clothing, especially silk, nylon, and rayon, while working in the processing room can also build up a static charge. Static electricity can also be built up rapidly in areas of low humidity, particularly during winter months when room moisture is normally low.

The three types of static markings commonly found on radiographs are *tree static, crown static,* and *smudge static.* Very rapid motions such as removing a film from interleaving paper can build up a charge sufficient to discharge in the form of tree static. Crown static is most often the result of the rapid withdrawing of a film from a tight or new box of film. Smudge static occurs when a discharge follows a path induced by dust, lint, or a rough intensifying-screen surface or work surface. Figure 2-10 illustrates the different types of static. Figure 2-11 shows static marks on exposed film.

While grounding the loading bench provides the simplest solution to static problems, moisture in the air around the film area is the greatest aid in the prevention of static. The higher the moisture content of the air, the less chance

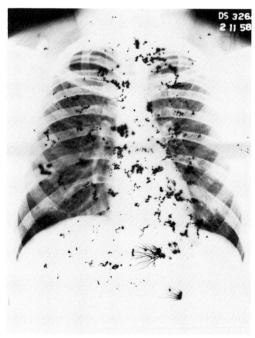

Figure 2-11. Static marks. A combination of smudge and tree static marks is shown.

there will be for static to appear. This is the reason why there is less static with manual processing than with automatic processing. Relative humidity ideally should be not lower than 45 percent, both for control of static electricity and film storage and for personal health.

Film identification data

In studying radiographs, the radiologist must be able to readily determine the side of the body examined, the name of the patient, the date of the examination, the direction of the central ray, and, if stereoradiography is employed, the direction of the tube shift. The most effective and accurate way to incorporate these data in the radiograph is by means of lead markers or other suitable radiopaque materials that can be placed on the cassette or film holder before the exposure is made. Letters, numerals, and holders for this purpose may be obtained from any x-ray supply vendor. Systems are also available to photographically flash this information on the film, such as the Kodak ID printer (Figure 2-12).

It is often necessary to identify on the radiograph the name and address of the radiologist or institution, as well as the patient's name, the case number,

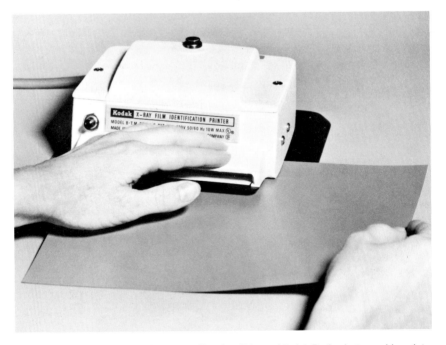

Figure 2-12. Kodak X-Ray Film Identification Printer, Model B. A photographic printer for including typed identification information on x-ray sheet film. The printing time is automatic for most medical x-ray films; a push-button switch increases the light intensity when necessary for other films. Separate slots are provided for inserting card and film. The printer operates on standard voltage and uses a 7-watt bulb. (Courtesy, Eastman Kodak Company, Rochester, New York.)

and the date of the examination. For the sake of appearance, efficiency, and medicolegal precaution, this information should be recorded permanently on all radiographs. Hand-written identification of a patient, placed on the radiograph after the film has been exposed and processed, is not admissible in a court of law.

BASIC FILM SENSITOMETRY

When one evaluates an x-ray film, normally a comparison is made with another brand of x-ray film whose properties are already known. If an accurate comparison between films is to be made—for example, to decide which would be better to use for a particular purpose—quantitative values must be assigned to the films. This is accomplished by using sensitometry.

THE SENSITOMETRIC CURVE

Sensitometry is defined as the quantitative measurement of the response of a film to exposure and development. The heart of sensitometry is the sensitometric strip, or step-wedge exposure (see below, under Testing X-Ray Film). A sensitometric strip is made by exposing successive areas on a film starting with a "no exposure" area and progressively making larger exposures until the maximum

Figure 2-13. The Cronex sensitometer. This sensitometer flashes reproducible simulated calcium tungstate exposures on both sides of the film. It is light in weight and portable. (Courtesy, E. I. du Pont de Nemours & Company, Wilmington, Delaware.)

density of the film is reached. These exposures are made in such a way that the amount of radiation to produce each step can be accurately measured. In a research laboratory, the successive steps are made with accurately calibrated radiographic equipment or a sophisticated sensitometer. A sensitometer (Figure 2-13) is a device that produces a constant simulated x-ray exposure. In the radiology department, the sensitometric strip exposure is made by making an x-ray exposure of a calibrated step wedge or with a calibrated sensitometer. A step wedge is so constructed that each level or step is relative to the next by a factor of $\sqrt{2}$. Thus each successive step on the sensitometric strip receives $\sqrt{2}$ or 41 percent more radiation exposure than the preceding step. Depending upon the situation, either seven or eleven successive exposures are made of the film. Figure 2-14 shows an aluminum and a plastic step wedge.

After the film is processed, the different degrees of blackening, i.e., the photographic densities, are measured with a densitometer. A densitometer is a device that measures the percentage of light that is transmitted through a film; the light transmittance is recorded in density units (Figure 2-15). A density unit is the logarithm of the reciprocal of light transmitted through a film and is expressed thus: density equals log incident light over transmitted light or

$$D = \log \frac{I_o \text{ (incident light)}}{I_t \text{ (transmitted light)}}$$

A density unit of zero would mean that 100 percent of the incident light has been transmitted through the film; a density unit of 1 means that 10 percent of the incident light has been transmitted through the film; a density unit of 3 would indicate that 0.1 percent of the incident light has been transmitted through the film. After the density units are read from each sensitometric step on the film and recorded, a graph is drawn using these values (see Figure 2-16). In the graph, the horizontal axis represents the sequence of steps in the step wedge;

Figure 2-14. At the left, a tissue-equivalent step wedge; at the right, an aluminum step wedge. Note the spin top attached to the lower portion of the aluminum step wedge.

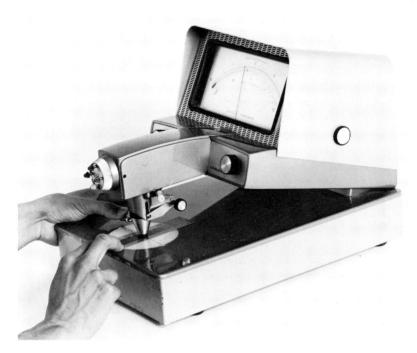

Figure 2-15. In this instrument, a light beam of constant intensity and hue is passed through the film and detected by the densitometer. Since the light source is constant, the amount of light that is transmitted through the film can be detected and displayed on the meter in terms of light transmittance or density units.

the vertical axis represents density units. A series of points is thus plotted, each representing the density record for a step of the step wedge. The points so plotted are joined together by a smooth curve, and thus a "characteristic curve" of the film is obtained. We can evaluate the characteristics of a film by simply noting the shape of the curve and its location on the graph.

Thus the product of sensitometry is a *sensitometric curve,* also known as a film characteristic curve or H & D curve (after Hurter and Driffield). A sensitometric curve is often called a "characteristic" curve since with it one can determine the basic characteristic of the film and its response to different exposure and processing conditions.

It must be emphasized that in order to have valid results and to compare one film with another, exposure and development conditions must be exactly the same for both. The characteristic curve, its slope, and its horizontal position relative to the axis will vary considerably according to the conditions of the film processing, the exposure technique factors and subjectivity of the reviewer. So many variations are possible in the factors that enter into sensitometry—the exposure time, mA, and kV, the chemistry-time-temperature factors of development, the special limitations of a particular densitometer, and the subjective bias of an interpreter—that it is impractical to compare one set of sensitometric determinations with results at a different time or at a different institution.

The characteristic curve and its interpretation are much like a patient's temperature chart, which is also a graph with a horizontal and a vertical axis. In a temperature chart the hours of the day are set along the horizontal axis and the patient's temperature is set along the vertical axis. When the temperature readings have been plotted on the chart and are joined with a line, one has a temperature curve. It is not necessary to examine each individual point on the graph; a glance at the complete curve tells one instantly how the patient is progressing. It is much the same with the characteristic curve.

The two basic factors that affect the sensitometric curve are (1) exposure factors, and (2) processing conditions. Each specific type of film has various characteristics, only one of which is the sensitivity (speed) to an exposure from

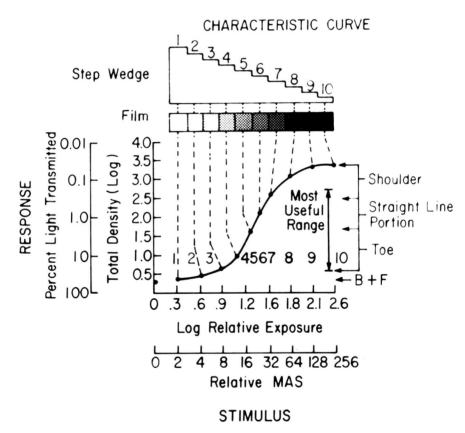

Figure 2-16. Components of a film characteristic curve. The step wedge shown at the top of the illustration allows exposures of the film which vary from each other by the square root of 2 (1.419) or 50 percent. The thickest part of the step wedge absorbs the most radiation and so allows the least exposure on the film; the thinnest part allows the greatest exposure. The amount of transmitted light is measured from the sensitometer strip and plotted in terms of density units along the vertical scale against the number of the step wedge along the horizontal scale. The components of the characteristic curve are as shown.

light or x-rays, as will be explained later. The sensitivity of the film or the efficiency with which the film responds to exposure is determined by the amount of exposure that is necessary to produce a certain density after controlled processing. If the film has a high sensitivity or speed, then (compared with a standard) for a given degree of density the exposure time will be small, and the film is referred to as being faster or having a higher speed. The sensitivity of the film to light or x-rays determines in turn the density of the finished product.

If exposure conditions are standardized, the conditions under which the radiograph is processed will considerably affect the sensitometric response of film. Raising or lowering of processing temperatures will increase or decrease the film speed, contrast, and base fog level of the film. Similar effects can be achieved by altering film immersion time in the developer. Changing the chemical activity —e.g., by substituting different brands of chemistry—may also do the same.

When performing film sensitometry one must use the same chemicals and the same processor and process the comparative films at the same time to keep variables to a minimum. Otherwise, sensitometric results may be questionable. Thus, it is the *relationships between* the exposure, processing, and inherent density qualities of the film that are ascertained by the measurement of the sensitivity of the film, that is by sensitometry. Those properties which influence the relationships between exposure, processing, and density are known as the sensitometric properties.

To interpret a sensitometric curve there are certain basic sensitometric properties which one needs to know.

Base density

The base density is the density resulting from the manufacture of the film and is inherent in the film base. In order to determine base density, the emulsion must be removed from the base. Soak the film in household chlorine bleach and simply wash the emulsion away. What remains is the film base itself, the density of which can be read with a densitometer. An average base density is 0.14.

Base plus fog

In the radiology department, one usually considers the combination of the inherent base density plus the fog produced by processing unexposed film. The fog itself is an inherent fog caused by the processing conditions. In practice, an acceptable base-plus-fog density for fresh film is approximately 0.18, which means that the inherent chemical fog is about 0.04.

D-min

The minimum density or D-min is the least density on the film after a proper exposure. The D-min represents the toe of the characteristic curve and is usually slightly higher than base-plus-fog density and is read from the first step of a sensitometric exposure. If the sensitometric strip is overexposed, the first step can be well above the toe; if underexposed, the toe may start at the third or fourth step.

D-max

The maximum density or D-max is obtained from the shoulder of the sensitometric curve and for practical purposes is read at the midpoint of the shoulder of the curve. True D-max is calculated from a white-light exposure and represents the maximum density available within a given emulsion.

Average gradient

The average gradient reflects the film contrast and is normally measured from points on the sensitometric curve which are 0.25 density units above base plus fog and 2.00 density units above base plus fog. Mathematically, average gradient is the slope of the line connecting these two density points. The average gradient is important because it includes the densities accepted as being the most useful in radiography. It includes some of the toe and generally the straight-line portion of the sensitometric curve. The more vertical this line, the greater is the film contrast (Figure 2-17).

A simple way to calculate the average gradient is to draw the line PR perpendicular from the first density point P, in Figure 2-17, parallel to the vertical axis, and another line QR to the right from point Q, parallel to the horizontal axis.

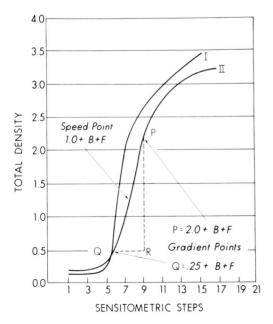

Figure 2-17. Film characteristic curves. The curve of film I is located to the left of that of film II and therefore film I is faster. In addition, film has a straighter slope than film II and therefore has higher contrast. The speed point and gradient points are noted. B + F is the base density plus fog.

Then one can measure the vertical line PR in millimeters and the horizontal line QR in millimeters. Dividing PR by QR will give the slope of the curve, which is the average gradient. For this procedure, horizontal units must be scaled to match vertical units. Otherwise, if the brand of graph paper is changed, the average gradient will also change.

Speed point

The speed point is also known as film sensitivity. The speed point is indicated by the position of the sensitometric curve along the horizontal exposure scale. Since most characteristic curves are plotted using log relative exposures, this exposure factor must be known before one film sensitivity can be compared with another. The speed point is the point at which a certain amount of exposure is required to produce a net density of 1.0 plus base plus fog. The more the speed point shifts horizontally to the left, the more sensitive or faster the film is, and vice versa (Figure 2-17).

Sensitometric curves are useful in comparing one film with another and in determining the effects of one type of chemistry on different types of films or vice versa. They are devices for controlling or monitoring film quality in the x-ray department from one day to another. They are also frequently used to match processors within the department so that an exposed film can be processed in any processor in the department without fear of adverse changes in the film quality.

The reciprocity law

From the preceding paragraphs it is evident that radiographic film density—the response of film to radiation—depends on a combination of the intensity of radiation and the duration of exposure of a film to radiation. This is a basic photographic principle advanced by Bunsen and Roscoe in 1875. Thus, an exposure can be mathematically defined as $E = It$ where E stands for exposure, I for intensity and t for the time of the exposure. This is commonly referred to as the *reciprocity law*. For all direct film exposure, this law holds true; the photographic effect is the same whether time is held constant and intensity is doubled or vice versa. One can then expect similar film densities from any particular combination of milliamperage (mA) and time (sec) as long as mAs remains constant and a *direct x-ray exposure* is made.

However, most x-ray film exposure is produced by the light photons released from the intensifying screens rather than by direct exposures. Oftentimes, a failure will occur of the reciprocal relationship between the intensity of the exposure and the duration of the exposure. This is called *film reciprocity failure*. For most radiographic procedures failure of the reciprocity law is of no significance. However, in very short exposures of the chest, such as those made at 10 ms or less, or for very long exposures, such as those for tomographic studies, reciprocity failure may be of some significance, and the technologist should be aware of this potential problem.

Failure of the reciprocity law can easily be detected by using the inverse square law (discussed in Chapter 7). If the x-ray equipment is properly calibrated, an exposure made at 180 cm (72 in.) with a 2-sec duration, should equal one made at 90 cm (36 in.) with a 0.5-sec exposure, choosing an appropriate kilovoltage and milliamperage which will allow a useful film density on a step wedge (see Figure 2-18). If the two film densities do not match, then one is left with the problem of determining whether film reciprocity failure exists or whether the equipment is not in calibration. If film reciprocity failure is suspected, more exposures using different time factors in the longer and shorter exposures must be made. Film speed can be determined from each exposure and plotted against time. From this data one can determine whether significant reciprocity failure is present. As a general rule, a film which does not exhibit any significant short-time reciprocity failure will have some degree of long-time failure and vice versa. The significance of this is that it is difficult to choose a film for chest radiography where exposure times are in the range of 10–15 ms and then to use the same film for a polytomographic study where exposure times may be in the range of 6 sec.

Figure 2-18. Step-wedge exposure made at 183 cm (72 in.), 2 sec (left), and at 91 cm (36 in.), 0.5 sec. (right). If the equipment is calibrated and no reciprocity failure exists, the two exposures should be essentially identical. However, in this case there is a full step of difference in density.

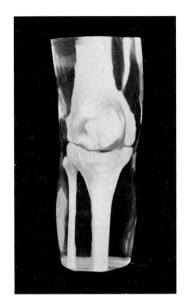

Figure 2-19. Phantom of the knee joint: human bones embedded in a special plastic.

TESTING X-RAY FILM

Every radiologic technologist should learn to make film tests to compare brands of film for contrast and speed values. He should also learn to run basic tests on processing chemicals. Tests should be made, with a step wedge, with sensitometric exposures, and with phantoms, for each brand or type of film and each brand or type of chemical to be tested. It is essential to use the same cassette and the same subject matter and to process all the films at the same time. Films should be tested at low and high kilovoltage.* All test exposure films should be filed for future reference in order to evaluate and compare at a later date the contrast and speed characteristics of a film emulsion with newer films and chemistry.

Patients should not be used for initial evaluation or comparison of film and chemistries. Good phantoms are available; it is needless to expose patients unnecessarily to radiation. The phantoms shown in Figures 2-19 and 2-20 are similar to ones used in most x-ray teaching laboratories and consist of carefully selected adult human bones, embedded in transparent, nongranular plastic. This plastic has absorption and secondary-radiation-emitting characteristics resembling those of living tissues. One may repeat an experiment, regardless of dosage, as often as necessary with no danger of excess radiation of a patient. One may also make critical detailed studies of bone-structure and sharpness comparisons.

---

* $\sqrt{2}$ step wedges are normally calibrated for use at a specific kilovoltage. When kilovoltage is changed, there is no longer a $\sqrt{2}$ relationship between steps.

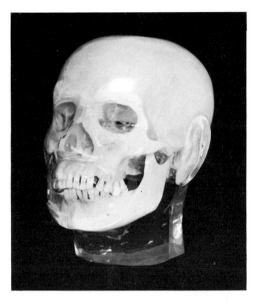
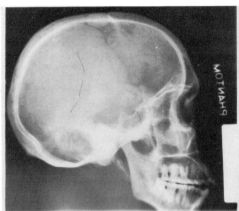

Figure 2-20. On the left is a human skull embedded in plastic. On the right is a radiograph of the phantom. The plastic serves to scatter and to absorb radiation so that a radiograph of the phantom is almost identical to exposures of human beings. Phantoms like this one and the one in Figure 2-19 are essential for teaching and evaluating film if exposure of a human subject to radiation is to be avoided.

The step wedge

The tissue-equivalent plastic step wedge (Figure 2-14) is an excellent tool for research and comparative studies. It consists of ten blocks which can easily be disassembled and rearranged to obtain different thicknesses. Embedded in the top are several test objects: a 2 × 2 × 2-cm lead plate for measuring magnification, distortion, and secondary radiation; five parallel pieces of stainless steel wire to determine sharpness; a cube of human bone; a gelatin capsule, 0.5 cm in diameter, containing bone dust; and an identical capsule containing air. The steps are wide enough to produce density strips on a radiograph which can be measured with a densitometer. Microdensitometer resolution measurements are possible across the shadow of the stainless steel wire. Densitometer measurements can also be made of images cast by other test objects. The plastic step wedge is large enough to be placed across a cassette containing several different screens and film, thus permitting screen and film evaluations with a single exposure. These absorption and secondary-radiation-emission characteristics of the wedge are similar to those of human tissues in the kilovoltage range between 50 and 150. Such a wedge is valuable in teaching and testing. One may demonstrate all the factors which control and affect the quality of the radiograph. For example, take a 35 × 43-cm (14 × 17-in.) film cassette with medium-speed calcium tungstate screens. Place an 18 × 43-cm (7 × 17-in.) comparative film in one half of the cassette and the test film in the other half.

Make a 70 kV, 100 mA, 100 ms exposure (single phase) of the step wedge, centering the step wedge longitudinally so that half of the step wedge covers half of each side of the test and comparative films. Process the films side by side and then plot sensitometric curves for both films and compare.

Care should be taken that the processor is clean and properly adjusted and that the output of the radiographic equipment is constant.

Before leaving this short summary of test procedures, it must be emphasized that films should be processed according to the manufacturer's explicit directions!

THE POLAROID RADIOGRAPHIC PROCESS

The Polaroid system consists of the radiographic packet, the radiographic cassette (which contains the energy-converting screen), and the film processor.

The Polaroid radiographic packet

The three primary parts (Figure 2-21) of the Polaroid radiographic packet are (a) the negative, which carries the photosensitive emulsion, (b) the pod, a metallic foil envelope which contains the viscous processing fluid, and (c) the receiving sheet, which yields the positive image. The remaining components permit the negative, the receiving sheet, and the viscous processing fluid to perform their collective function of producing a finished radiograph. The radiograph obtained

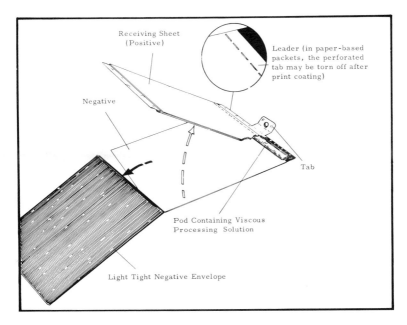

Figure 2-21. The Polaroid radiographic packet.

can be a paper-based reflection print (type 3000X) or a polyester-based translucency (type TLX film).

With TLX film, the final result is a translucent positive radiograph, which can be viewed either by transmitted light, like that from a standard illuminator, or by reflected light, like that available from room illumination.

Type 3000X yields a positive radiographic print which can be viewed by reflected light only. When Polaroid type 3000X or Polaroid type TLX packets are exposed to x-radiation in conjunction with a blue-emitting, high-speed fluorescent screen, radiographs of satisfactory density are produced.

Exposure

Figure 2-22 represents how a Polaroid radiographic packet is exposed. The fluorescent screen emits visible light when excited by x-rays. This visible image is recorded as a latent image on the photosensitive surface of the negative.

Processing

The exposed Polaroid radiographic packet with its latent image is processed by drawing it through a set of pressure rollers which burst the pod and spread

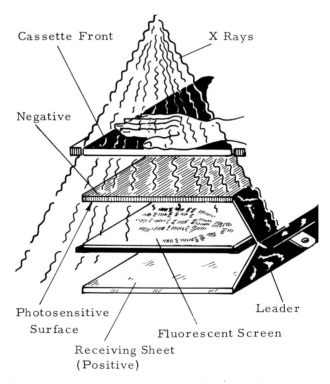

Figure 2-22. The exposure of a Polaroid radiographic packet.

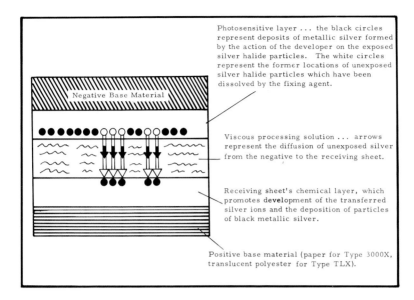

Photosensitive layer ... the black circles represent deposits of metallic silver formed by the action of the developer on the exposed silver halide particles. The white circles represent the former locations of unexposed silver halide particles which have been dissolved by the fixing agent.

Negative Base Material

Viscous processing solution ... arrows represent the diffusion of unexposed silver from the negative to the receiving sheet.

Receiving sheet's chemical layer, which promotes development of the transferred silver ions and the deposition of particles of black metallic silver.

Positive base material (paper for Type 3000X, translucent polyester for Type TLX).

Figure 2-23. The process of diffusion in Polaroid film, shown schematically.

the viscous processing fluid between the negative and the receiving sheet. The developing agent in the processing fluid reacts with *exposed* silver halide grains and forms a visible negative image by reducing the silver ions to metallic silver. While the negative image forms, a silver ion complexing agent forms a complex of very high solubility with the *unexposed* silver halide grains. The solution of complex ions diffuses through the thin layer of viscous processing fluid to the receiving sheet, where, in contact with the imaging elements, the soluble silver ion complex is reduced, forming metallic silver and thus yielding the positive imaging. This entire diffusion transfer reversal process is accomplished in 10 seconds for 3000X film and in 45 seconds with type TLX. Figure 2-23 shows the process schematically. The width of the layers has been exaggerated in this illustration for clarity.

One can perhaps better understand why the final radiograph bears a positive image (see Figure 2-24) by recalling that it is the unexposed silver halide grains that aré the source of the metallic silver deposits which form the positive radiographic image.

Coating

Within one hour after processing, Polaroid radiographic prints should be treated with a print-coating solution that is provided with the film and according to the instructions. Coating of the film quickly neutralizes the small amount of processing solution that remains in the receiving sheet when it is stripped from the negative. Moreover, as soon as the print-coating solution dries, it leaves behind a durable plastic film which protects the surface of the radiograph from scratches, moisture, and atmospheric contaminants which can attack the silver image.

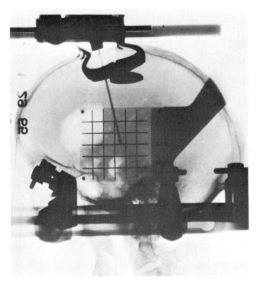 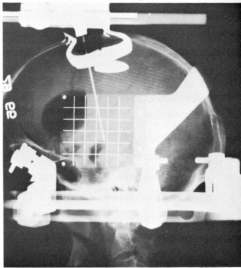

Figure 2-24. At the left, a radiograph of lateral skull made with Polaroid TLX film—a positive. The metal clamps show up dense black. At the right, a radiograph of the same skull made with conventional radiographic film—a negative.

XERORADIOGRAPHY

The implementation of mammography as a modality for breast cancer detection brought with it a new imaging technique for the radiologic technologist. Although the process of xeroradiography has been used for a number of years, it was the work of Dr. John Wolfe, Radiologist at Hutzel Hospital in Detroit, that brought it to the forefront and made it a competitor to the film format. Today, in addition to its applications in mammography, xeroradiography is used in examining small anatomical parts. Because of the lower speed of xeroradiography as compared with a medium-speed film/screen combination, it is generally not used for radiographing larger anatomical parts.

Almost all students are familiar with Xerox machines. Xeroradiography is the combination of the xerographic copying technology and conventional x-ray exposure techniques using standard radiographic equipment. However, the development of a latent and visible image is entirely different from that of conventional radiographic images.

The Xerox 125 system is used in xeroradiography. It consists of two components: a conditioning unit and a processing unit. The purpose of the conditioning unit is to remove residual electrical charges on a selenium plate and then to reapply a uniform charge across the plate. The processing unit converts the latent image to a visible image.

The xeroradiographic image

In xeroradiography the image receptor—any device which receives an image is called an image receptor—is an aluminum plate or substrate coated with a thin layer of selenium alloy (Figure 2-25). Selenium is a photoconductor; that is, selenium changes its resistance to electrical flow (its electrical conductivity) in proportion to the intensity of light or x-rays which it absorbs. Photoconductivity then is electrical terminology for the property shown by certain materials of varying their resistance in proportion to illumination. Photoconductivity is the basic electrical property of a television vidicon camera tube.

Prior to exposure, the selenium-coated plate must be made receptive to an image. This is accomplished by placing the selenium plate under an ion-emitting device called a *scorotron* (Figure 2-26) which consists of a series of fine screen wires. When a positive voltage is applied to the scorotron, positively charged air ions are created, which are then uniformly attracted to the selenium surface. At this point the positively charged selenium plate is light sensitive and is similar in light receptivity to the silver halide crystals in a film emulsion. Needless to

A.

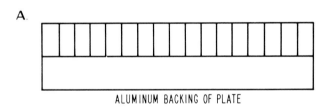

ALUMINUM BACKING OF PLATE

B.

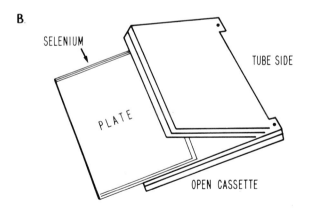

SELENIUM

TUBE SIDE

PLATE

OPEN CASSETTE

Figure 2-25. A. Cross section of a selenium plate. **B.** A plastic cassette houses the charged plate to protect it from light from the time of initial charging until it is inserted into the xeroradiographic processor. (Redrawn from *Technologist's Guide to Detection of Early Breast Cancer by Mammography, Thermography, and Xerora-diography,* American College of Radiology, Chicago, Illinois.)

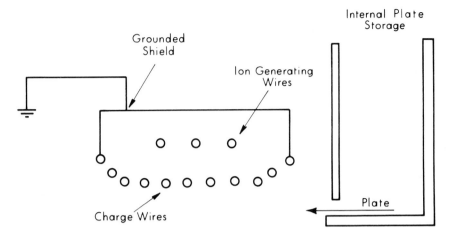

Figure 2-26. Pictorial presentation of a scorotron, a device which applies a uniform charge across the plate. (Redrawn from *Technologist's Guide to Detection of Early Breast Cancer by Mammography, Thermography, and Xeroradiography,* American College of Radiology, Chicago, Ilinois.)

say, this positive ion charge must be applied to the selenium just prior to the x-ray exposure or it will dissipate. The amount of voltage differential applied to the selenium plate will in part determine the contrast level of the developed image.

The conditioning unit works in this fashion (Figure 2-27). As a storage box containing the selenium plate is placed in the unit, the selenium plate is removed and transferred through a heating oven, where any residual electrical charge is removed from the plate, and from there to a storage area. When an empty cassette is placed in the receiving area, a selenium plate is mechanically moved from the storage area to the scorotron, where it receives its electrical charge and then enters a light-tight cassette. The cassette is then ready to be used.

Formation of the xeroradiographic latent image

When the cassette containing the ion-charged selenium plate is struck by radiation, the x-ray photons penetrating the anatomical part are absorbed by the selenium and create ion pairs which discharge or neutralize the positive ion charge on the selenium plate in proportion to the amount of radiation striking the plate (see Figure 2-28). The remaining positive charges and neutral areas on the plate form a latent image similar to the latent image formed on film when using conventional radiographic techniques, except that in xeroradiography the latent image is composed of an ion charge rather than sensitized silver halide crystals.

At this point the xeroradiographic latent image is pressure sensitive, just as conventional x-ray film is light sensitive. The selenium plate should be processed as soon as possible to prevent dissipation of the ion image (latent image failure).

Developing a visible radiographic image

In the Xerox 125 system, the latent image is developed through the other component of the system, the *processor* (Figure 2-27). After the cassette enters the processor, the selenium plate is mechanically removed from the cassette and moved into a development chamber. A uniform positive charge is applied to the aluminium backing of the plate. During the process of development, a cloud of charged powder particles (toner) is released. The negative particles in the cloud are attracted to the selenium plate, where they are deposited in proportion to the amount of charge remaining on the plate. The amount of toner deposited is controlled by the residual latent image charge on the selenium plate and the electrical fields which exist during development. The deposition of toner on the selenium plate produces a visible image which is transferred to a paper mask for convenience in handling and interpretation. After the image has been transferred to the paper sheet, the selenium plate is shuttled to another area in the processor where residual toner is removed, first by exposure to light and then by mechanical cleaning. The plate is then ready for re-use.

Since the greater the number of x-ray photons that strike an area on the selenium plate, the greater the cancellation of the ion charge, it follows that the

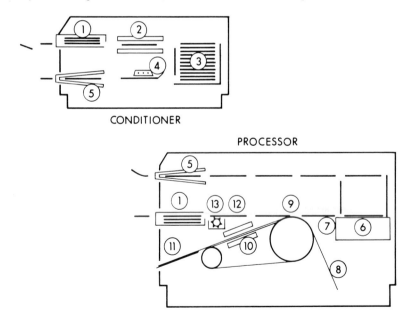

Figure 2-27. An internal sketch of the conditioner and processor. **1.** Storage box station. **2.** Relaxation oven. **3.** Storage magazine. **4.** Plate charge device. **5.** Cassette station. **6.** Development chamber. **7.** Pretransfer. **8.** Paper feeder. **9.** Transfer. **10.** Fusing oven. **11.** Output tray. **12.** Precleaning. **13.** Plate cleaning. (Redrawn from *Technologist's Guide to Detection of Early Breast Cancer by Mammography, Thermography, and Xeroradiography,* American College of Radiology, Chicago, Illinois.)

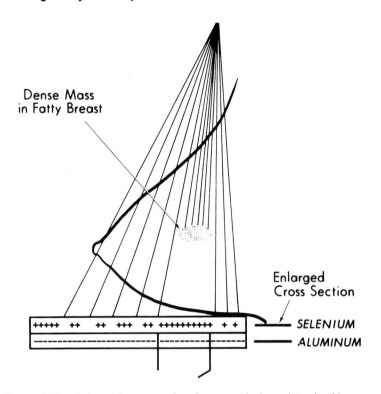

Dense Mass
in Fatty Breast

Enlarged
Cross Section

SELENIUM

ALUMINUM

Figure 2-28. A latent image produced on a selenium plate. In this case, x-rays pass through a breast. Some of the photons are absorbed and others pass through the breast tissue and strike the selenium plate. The positive charge on the selenium plate is selectively dissipated in proportion to the incident radiation. An electrical charge is formed on the plate that corresponds in all aspects to the x-ray image that is present on the plate. This electrical latent image is held on the selenium plate by the negative charge of the aluminum backing. Forming an image in this manner is called electrostatic imaging. (Redrawn from *Technologist's Guide to Detection of Early Breast Cancer by Mammography, Thermography, and Xeroradiography,* American College of Radiography, Chicago, Illinois.)

less the amount of charge, the less the amount of toner attracted to the point on the selenium plate; thus the resultant visible image is a positive rather than a negative. When a negative is desired, the polarity of the development process can be reversed by simply changing a switch position, and a negative image is formed. In positive development, tissues which absorb little radiation, such as fat, will appear as very light blues on the visible image, and heavy densities, such as thicker tissues or bone, will appear as dark blues.

Edge enhancement in xeroradiography

Inherent in the xeroradiographic process is a process called edge enhancement (Figure 2-29). Edge enhancement is an exaggeration of the actual contrast at

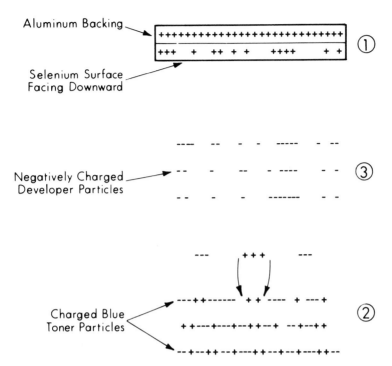

Figure 2-29. The principle of developing a xeroradiograph. An exposed selenium plate is positioned in the development chamber with the selenium surface of the plate facing downward. A cloud of negatively and positively charged blue toner is released into the chamber. The negatively charged toner particles are attracted to and deposited on the selenium plate by a large positive voltage applied to the back of the plate. The intermediate visible image is then transferred to a paper or plastic plate for permanent storage. (Redrawn from *Technologist's Guide to Detection of Early Breast Cancer by Mammography, Thermography, and Xeroradiography*, American College of Radiology, Chicago, Illinois.)

the borders of an object or density which allows one to easily perceive density differences. We cannot here go into the physics of how edge enhancement occurs. Suffice it to say that each positive ion that remains on the selenium plate after exposure carries with it a minute electrical force field, somewhat analogous to the magnetic field surrounding an electrical coil. The more ions there are, the stronger the force field, and this field becomes synergistic rather than additive. At the edge of two radiographic densities, there are more positive ions at the less exposed edge than there are on the more exposed edge. Similarly, the positive ion force field, because of the synergistic effect of the ions, is much stronger than one would normally expect. The next result of this force field is that a proportionately larger amount of toner is attracted to the edge of the opposing densities. This produces greater visibility of the edge of the densities. To say it in another way, edge enhancement occurs where there is an abrupt change in electrical charge, as the high side of the charge causes more toner to be deposited and the low side of the charge causes less toner to be deposited.

Broad object latitude of xeroradiography

Because of the electrostatic principle of xeroradiographic imaging, another feature of the process is that of wide object latitude. A wide object latitude implies the ability to visualize structures in a single image over a greater range of object thicknesses. For example, when using radiographic film and making an exposure of the hand, it is difficult to penetrate the wrist, the phalanges, and the soft tissues all equally, since some of the anatomical areas will absorb more radiation than other areas. The result may be properly exposed structures of the wrist, but overexposure of the soft tissues of the hand. In radiography, this can be overcome by using a long-latitude film or direct-exposure film. In xeroradiography a wide object latitude results from the edge enhancement feature plus a tendency to subdue gradual changes in object densities.

ELECTRONIC RADIOGRAPHY

In 1972 Xonics Corporation introduced another form of radiographic imaging called electronic radiography (ERG). Electronic radiography is an extension of ionography, a process of depositing an electrical charge on a plastic sheet rather than having it leaked off a selenium plate as in xeroradiography.

The electronic radiographic process requires four operational steps, just like conventional radiography: preparation, exposure, development, and fixing. In lieu of radiographic film, electron radiography uses a clean, clear polyester sheet similar to the conventional film base. This plastic sheet is mounted on a metal plate and all electrical charges are removed from the base/plate combination. The image receptor is placed in an imaging chamber, which is then filled with a gas. A high-potential electrical field is applied to the image chamber. This supplies a voltage differential for the image chamber and also ionizes the gas.

Exposure is performed in the usual manner. During the exposure (Figure 2-30), electrons are released in the chamber electrode and trigger a proportional release of electrons in the gas contained in the exposure chamber. Electrons released by the discharge are attracted to the positive or image-receptor side of the image chamber and are deposited on the plastic sheet in direct proportion to the x-ray energy. Hence, an electrical charge is developed on the plastic sheet which corresponds in fidelity to the original x-ray image and represents a latent image just as in conventional film.

Development of the latent image is accomplished by automatically spreading an opaque liquid or powder in the image chamber. The voltage charge on the plastic sheet attracts particles of this developer; more particles are deposited on high-voltage areas and fewer on low-voltage areas. Fixing of the image entails permanently bonding the developer to the plastic sheet by either a heat or a

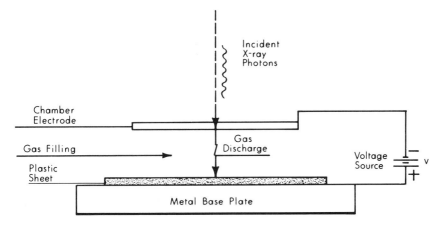

Figure 2-30. Principles of electronic radiography. The receptor of electronic radiography is composed of a chamber electrode electrically connected to a metal plate on which is attached a plastic sheet, which serves as the image receptor. In this type of electrostatic receptor, the incident x-ray photons release electrons in the gas-filled chamber which are deposited proportionately on the plastic sheet. The electrical latent image is converted to a visible image in a way similar to that of xeroradiography. In this type of electrostatic imaging, electrons are collected on an image receptor. In xeroradiography, an electrical charge on a selenium plate is dissipated, and the electronic image remains.

chemical process. The visible image is viewed as in a conventional film system.

Although electronic radiography is still in the research stage, it promises to offer a suitable image receptor in the future. Since the recording system does not use a silver halide emulsion, potential savings in film costs can be tremendous—but unfortunately, the cost of acquiring and maintaining the complex equipment could easily exceed film cost. Preliminary results also indicate that in exposure of the patient to radiation, it is comparable to a medium-speed film/screen combination and much better than the xeroradiography system. Electronic radiography is reported to offer some degree of edge enhancement, but not at the expense of decreasing adjacent density as in xeroradiography.

There are other forms of radiographic imaging which are more appropriately covered in other publications. These include ultrasound, holography, thermography, and computerized tomography. Though imaging in radiologic technology is changing rapidly, it is mandatory that all technologists be thoroughly familiar with the contents of this chapter as a strong foundation.

SUGGESTED READING

Christenson, E., Curry, T. S., and Nunnally, J. *An Introduction to the Physics of Diagnostic Radiology.* Lea and Febiger, Philadelphia, 1972.
Cullinan, J. E. *Illustrated Guide To X-Ray Techniques.* Lippincott, Philadelphia, 1972.

Darkroom Technique for Better Radiographs Processed Manually or Automatically. Du Pont, Wilmington, Del. (No date.)

Egan, R. L., ed. *Technologist's Guide to Detection of Early Breast Cancer by Mammography, Thermography, and Xeroradiography.* American College of Radiology, Chicago, 1976.

Principles of Subtraction in Radiology. Du Pont, Wilmington, Del. (No date.)

Radiographic Processing and Film Quality Control. American Society of Radiologic Technologists, Chicago, sixteen video tapes with accompanying study guide.

Seemann, H. E. *Physical and Photographic Principles of Medical Radiography.* Wiley, New York, 1968.

Staton, L., Brady, L. W., Szarko, F. L., Day, J. L., and Lightfoot, D. A. "Electron Radiography: A New X-Ray Imaging System." *Applied Radiology* 2, no. 2 (1973):53–56.

Weaver, K. E., Wagner, R. F., and Goodenough, eds. *Medical X-Ray Photo-optical Systems Evaluation.* DHEW Publication (FDA) 76–8020. October, 1975.

REVIEW QUESTIONS

1. Gelatin is chemically considered to be a

    a. Compound
    b. Mixture
    c. Solution
    d. Element
    e. Colloid

2. A processed radiographic film is a

    a. Composite
    b. Negative
    c. Positive
    d. Print
    e. Electrostatic negative

3. Modern x-ray film base is composed of

    a. Cellulose acetate
    b. Polyester
    c. Cellulose nitrate
    d. Nylon
    e. Polyacetate

4. A film base is normally how thick?

    a. 5 mils
    b. 7 mils
    c. 9 mils
    d. 12 mils
    e. 15 mils

5. "Crosstalk" is caused by

    a. Emulsions too close together
    b. Improper film/screen combination
    c. Oblique scattering of x-rays
    d. Focal spot too small
    e. Exposure caused by photons from opposite screen

6. The emulsion most commonly used in making medical x-ray film contains

   a. Silver nitrate
   b. Silver sulfite
   c. Silver bromide
   d. Silver fluoride
   e. Calcium tungstate

7. Film emulsion is composed of

   a. Potassium bromide
   b. Alum
   c. Metol plus gelatin
   d. Silver halide plus gelatin
   e. Gelatin alone

8. Film grain is the result of

   a. Clumping of silver halide particles during manufacturing
   b. Processing at too low a temperature
   c. Overactive phenidone
   d. Underactive hydroquinone
   e. Screen speed too fast

9. In medical radiography, a latent image is defined as

   a. A visible, unprocessed image
   b. An invisible image made visible by processing
   c. An unuseful image
   d. A photographic image
   e. An electrostatic image

10. Increasing the temperature at which a film is processed will

    a. Decrease the time required for development
    b. Increase the time required for fixing
    c. Have no effect on film
    d. Lower chemical activity
    e. Cause bromide retention

11. Screen type film is _____ when used with intensifying screens and _____ when exposed by direct radiation.

    a. Faster
    b. Slower
    c. No effect

12. Panchromatic film is defined as

    a. Primarily receptive to red
    b. Primarily receptive to green
    c. Primarily receptive to blacks and whites
    d. Primarily receptive to x-rays
    e. Receptive to all colors

13. Spot film or photospot camera film is available in all the sizes listed below, EXCEPT

    a. 50 mm
    b. 70 mm
    c. 90 mm
    d. 100 mm
    e. 105 mm

14. Subtraction film is primarily used in

    a. Angiography
    b. Mammography
    c. Routine radiography
    d. Osteology
    e. Urography

15. Film should be stored at

    a. 10–21°C
    b. 5–9°C
    c. 21–29°C
    d. 0–5°C
    e. None of the above

16. The phrase "first in, first out" means

    a. Cassettes should be used in the same order as they come out of the darkroom.
    b. The oldest film in stock should be used first.
    c. The first patient to arrive in the x-ray department is examined first.
    d. Freshest chemistry is used first.
    e. First film out of the processor is discarded.

17. Types of static markings commonly found on radiographs include

    a. Tree static
    b. Crown static
    c. Smudge static
    d. A, B, and C
    e. A and B only

18. The term "base plus fog" means

    a. Thickness of the base plus radiation fog.
    b. Fog level in a film prior to processing.
    c. Inherent density in the film base due to manufacturing process plus fog produced by spontaneous development of silver halide crystals, without exposure.
    d. Scattered radiation fog enhanced by processing.
    e. Inherent base film density plus scattered radiation.

19. On a standard sensitometric exposure, each radiographic step is approximately how much greater than the preceding step?

    a. 25%
    b. 40%
    c. 95%
    d. 100%
    e. 150%

20. Density in a radiograph is the

    a. Differences between blacks and whites
    b. Overall blackness of the radiograph
    c. Amount of distortion of the image
    d. Clearness and sharpness of the image
    e. Reciprocal of the amount of light transmitted

21. The most important factors which affect film sensitometry include

    a. Time of development

b. Temperature of developer
c. Activity of developer
d. A, B, and C
e. A and B alone

22. Average gradient refers to

    a. The slope of a curve
    b. Film contrast
    c. Film latitude
    d. A, B, and C
    e. A and B alone

23. Xeroradiography is used primarily in

    a. Angiography
    b. Mammography
    c. Routine radiography
    d. Osteology
    e. Urography

24. The image receptor in xeroradiography is

    a. A selenium plate
    b. An aluminum plate
    c. A plastic sheet
    d. An x-ray film
    e. An image intensifier

25. The image receptor in electronic radiography is

    a. A selenium plate
    b. An aluminum plate
    c. A plastic sheet
    d. An x-ray film
    e. Polaroid film

3 CASSETTES AND SCREENS

Cassettes and screens are extremely important in formulating radiographic techniques. They are a part of the total radiographic system. One part of the system cannot be ignored since all parts of the system play a role in the end product, a diagnostic radiograph.

FILM CASSETTES

A cassette is a thin, light-tight container slightly larger than the film it is intended to hold. The front is made of a radiolucent material such as Bakelite; the back is made of steel or of a lightweight metal such as magnesium. Felt gaskets are added to insure light-proof edges. The inner (back) side of the cassette (except in some phototimer cassettes) is lined with a thin layer of lead which absorbs secondary radiation emitted by the back of the cassette, Bucky tray, or tabletop. A good cassette is light-proof and provides uniform contact between film and screen over the entire area of the film. A bent or warped cassette front causes poor film/screen contact and results in blurring of the radiographic image.

Phototimer cassettes

The cassettes used for conventionally timed radiography have lead-lined backs to reduce the effect of backscattered radiation on the film. Cassettes used for phototiming have a radiolucent back to permit the radiation reaching the film to continue on the automatic exposure control (AEC). (Note: some phototiming cassettes have a very thin layer of lead to absorb scattered radiation; but most do not.) The AEC measures the amount of radiation reaching the film and terminates the exposure according to a predetermined density scale. If the automatic exposure control is located in front of the cassette, phototiming cassettes are not required. Care should be taken that phototiming cassettes are appropriately identified so that they are not mistaken for standard cassettes. Standard cassettes used in phototiming applications, where a radiolucent back is required, increase exposure time, with the result that the exposure is terminated after a proper density is achieved, and the outcome is an overexposed radiograph. Motion unsharpness may also occur when the exposure time is too long to provide stop-motion of movement occurring during the time of exposure.

PROBLEMS WITH CASSETTES

Numerous problems can occur with film cassettes, including bent or warped cassettes, light leaks, non-standardization, air trapping, and poor screen contact.

Light leaks

In most cassettes the felt gasket serves a dual purpose. It affords a light-tight seal preventing the enclosed film from being fogged around the edges by light leaks, and it also insures a better film/screen contact. Figure 3-1 shows film fogged from a light leak.

Poor screen contact

Poor screen contact results from a warped or bent cassette and is usually caused by misuse or dropping the cassette. On the finished radiograph poor screen contact presents itself as areas of poor definition or detail. Figure 3-2 shows a radiograph with poor definition due to a warped cassette front.

Poor screen contact may be transitory. A transitory poor screen contact can result from a cassette bending under the weight of a patient—for example, during portable examinations. In this situation the patient's body weight will bend the cassette, resulting in poor screen contact during the time of the exposure; but the cassette will spring back to its normal shape after it is removed from under

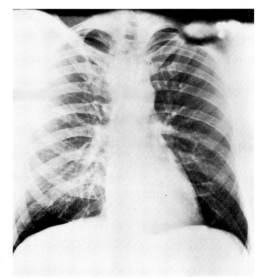

Figure 3-1. The black density at the top of the film is caused by a light leak at the top of the cassette.

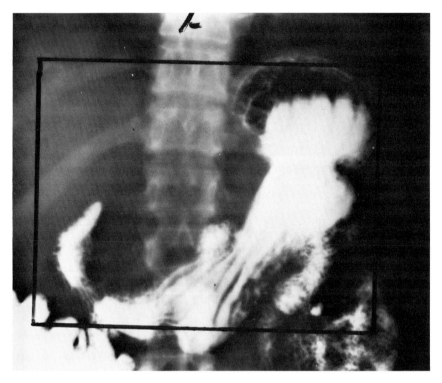

Figure 3-2. Poor definition due to a warped 14 × 17 in. cassette front. Notice the lack of sharpness in the outlined area. The figure has been enlarged to better demonstrate the area.

the patient. This situation can be corrected by selecting rigidly constructed cassettes for portable examinations.

To test for screen contact, lay a wire screen (about $\frac{3}{16}$-in. mesh) on top of the cassette to be tested. Make an exposure at 50 kV, 50 ms (0.05 sec), 50 mA, 90 cm (36 in.) source/image-receptor distance (SID), single-phase equipment. Develop the film in the usual manner. If the radiograph shows areas of blurring or fuzziness of the wire mesh, the contact between screen and film is probably poor. It is wise to check the front of the cassette before removing the screens, since it may be warped; or the back of the cassette may be sprung at the hinged area because the cassette has been dropped. Figure 3-3 shows poor contact due to a warped front before and after repairs. Cassettes should be repaired if at all possible, since repairs will result in substantial savings and the screens will not have to be replaced. It is false economy to mount new screens in faulty cassettes.

Other cassette faults which affect film quality include worn felt, loose hinges, dented fronts, and sprung frames. Although cassettes are sturdy and strong enough to withstand normal hard use, they cannot withstand abuse.

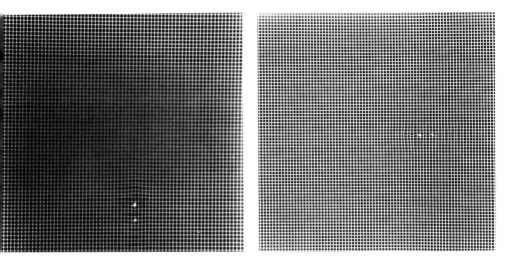

Figure 3-3. Screen wire mesh test for screen contact of a 14 × 17 in. cassette. (Left) In the center of the illustration notice the lack of sharpness of the wire mesh as compared with the wire mesh screen on the edges of the illustration. (Right) The same cassette after repair. Notice how uniformly sharp the wire mesh is.

Air trapping

Most radiographers are not aware of the problem with air trapping, which produces a transitory poor film/screen contact. Air trapping occurs when the film is inserted in the cassette and the cover rapidly closed. This rapid movement causes air to be trapped in the cassette, separating the film from the screen.

Figure 3-4. A film cassette with a curved back. When the cassette is closed, air is expelled; this eliminates air trapping and transitory poor screen contact. (Courtesy, E. I. du Pont de Nemours & Company, Wilmington, Delaware.)

65

These pockets of air will gradually dissipate, but the time required for this to happen is from minutes to hours, depending upon the construction of the cassette. If a cassette is put in use shortly after the film is loaded, air trapping can be a significant problem.

Air trapping can be avoided if care is taken in loading the cassette. To counter the problems, several manufacturers have introduced cassettes (Figure 3-4) which are rigidly constructed, but have a curved back and/or front which simply rolls the air out of the cassette as the cover is closed.

Incidentally, air trapping can cause a significant problem in cut-film serial film changers. With the older type of film changers, the screens were flat and trapped air as they closed. It requires approximately 15 ms for this air to escape, and this time must be allowed for in establishing film techniques. Some newer film changers have curved screens which minimize this problem.

INTENSIFYING SCREENS

Intensifying screens are used in radiography to enhance the effects of radiation in producing film blackening and to increase contrast. Screen technology has been at a standstill for many years, and only lately have great advances been made. The selection of an intensifying screen can make or break the production of a good technological examination.

In 1896, only one year after Roentgen discovered x-rays, Thomas Edison found that scheelite, commonly known as calcium tungstate, a naturally occurring mineral, fluoresced a bright blue color on exposure to the new-found rays. His discovery was not an accident, but the result of a thorough systematic examination of no fewer than one thousand heavy-metal salts. Calcium tungstate became the major phosphor used in the production of medical x-ray intensifying screens until the introduction of "rare earth" screens.

Construction of an intensifying screen

Figure 3-5 shows the cross section of a radiographic intensifying screen. The screen is composed of an abrasive or protective coat, a phosphor layer, a reflective coat, and a polyester (plastic) or paper base.

The function of the protective coat is to prevent damage to the phosphor, to prevent moisture from entering the phosphor layer, and to provide a cleanable surface. Some of the newer phosphors, such as cesium iodide used in image intensifiers, are very hygroscopic (i.e., absorb water from the air) and for this reason are impractical for use in intensifying screens. The texture of the protective coat becomes very important—particularly in film changers, where the film is liable to be scratched as it moves across them.

The radiation reduction effect of intensifying screens lies in the next component,

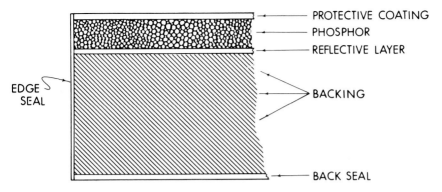

Figure 3-5. Cross section of an intensifying screen.

the phosphor layer. Certain substances have the ability to emit light after they are irradiated. These are called phosphors. (The word *phosphor* comes from the Greek word *phosphoros,* meaning "light bringer" or "light bearer.") In radiological terminology, a phosphor converts radiant energy into light energy. The phosphor layer is composed of small, heavy-metal crystals dispersed in a solvent and binder. In manufacturing the screens, every effort must be made to have all the crystals of the same size and evenly dispersed in the emulsion. The purpose of the binder is to hold the crystals in place. Dyes (activators) can be added to the binder to shift the natural blue spectral emission of the calcium tungstate crystal towards the yellow or green spectrum, slowing the screen speed and thus improving detail and controlling the lateral dispersion of light.

The dispersion of the crystals in the phosphor layer produces what is known as *screen grain* or *screen mottle.* Figure 3-6 (right view) is an electron micrograph of a phosphor which shows uniform dispersion of the tungstate crystals, and hence very little screen mottle or grain. Figure 3-6 (left view) shows a clumping

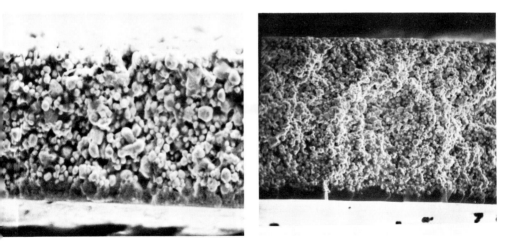

Figure 3-6. (Left) Photomicrograph showing clumping of screen phosphor. (Right) Screen phosphor, showing an even dispersal of phosphor in the binder. Screens with phosphor clumping will have objectionable screen mottle. (Courtesy, Minnesota Mining and Manufacturing Company, St. Paul, Minnesota.)

of the phosphor particles; screens made from this emulsion would have high screen mottle or grain.

The third component of a screen is the reflective layer. The purpose of the reflective layer, as its name implies, is to reflect the light generated by the phosphor towards the film. As an x-ray photon interacts with a crystal, light may be emitted in any direction. Since 99 percent of the x-ray energy is lost anyway, as much of this light as possible needs to be utilized, and the reflective layer achieves this; that is, it reflects the light emitted away from the film back toward the film.

Screen fluorescence

If the light output from an intensifying screen persists for no longer than a few microseconds ($< 10^{-8}$ sec), the screen is said to be *fluorescent.* If the screen light persists for longer than 10^{-8} sec, the screen is said to be *phosphorescent.* Phosphorescent screens are generally used in radiography. The term "screen lag" is used to describe an additional very long-lived component of the emissions. Screen lag can be measured in most commercial phosphors, but the intensity of the component is so low that it no longer has any practical significance.

Screen phosphors

The most commonly used phosphor in construction of medical x-ray intensifying screens is calcium tungstate. When calcium tungstate is struck by x-ray photons, it emits light photons with a broad spectrum of wavelengths centered in the region of 450 nanometers (4500 angstroms). This output is in the blue-light spectrum. Typical x-ray film is manufactured to be most sensitive to this wavelength.

In the new rare earth screens, phosphors are composed of either yttrium oxysulfide, gadolinium oxysulfide, barium fluorochloride, lanthanum oxybromide, lanthanum oxysulfide, or a mixture of these phosphors. The term *rare earth* is applied to these new screens, but not all of them are made from rare earth elements. Gadolinium, lanthanum, and yttrium are rare earth elements, falling in the Periodic Table under Group 3B (Group 3B is also called the Lanthanide Series), but barium is not. The spectral emission of some of the rare earth screens is considerably different from that of calcium tungstate. Whereas calcium tungstate emits a broad spectrum of light centered in the blue wavelengths, the emission for the true rare earth screens consists of very sharp peaks (Figure 3-7). Some of the rare earth screens, such as gadolinium oxysulfide, peak at a shorter wavelength and emit a light in the green-light spectrum. These screens can be used with blue-sensitive film; however, some loss of speed will be noticed. An activator can be added to the screen to shift the green spectral emission toward the blue spectrum. A specially made film which is green-light sensitive should be used with these screens requiring "green sensitive" film.

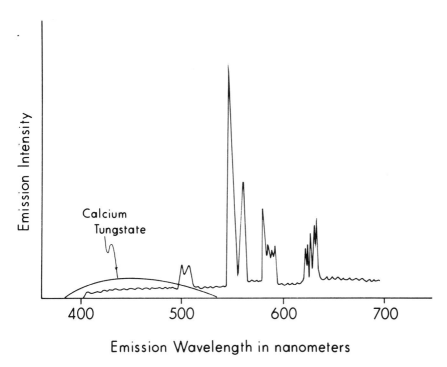

Emission Wavelength in nanometers

Figure 3-7. A comparison of the emissions of calcium tungstate and of rare earth screens. The emission of calcium tungstate screens is in a broad spectrum centered at 450 nm. Rare earth emissions are in high, sharp peaks, generally at longer wavelengths.

Screen efficiency

When an x-ray photon strikes a screen phosphor, events of a number of kinds are possible. Figure 3-8 is a simplified view of the screen imaging system which breaks down the various parameters which are collectively responsible for the efficiency or inefficiency of the screen. As noted in the figure caption, there is a finite probability that an incident x-ray photon will be absorbed by an intensifying screen and there is a finite probability that the absorbed x-ray energy will be converted into light. A finite probability exists that the converted light will escape from the screen in the right direction to hit the film, and finally there is a finite probability that the incident light will expose a latent image on the film.

A probability equal to 1 means that an event will occur 100 percent of the time; a probability of 0.5 means that the event will occur 50 percent of the time; and so on. To be totally efficient every photon being absorbed by a screen must produce a light photon, which in turn must contribute to the latent image formation. Unfortunately, this does not occur; the probability of every photon contributing to the latent image formation is very low, and thus the efficiency of screens is very low. Table 3-1 lists the efficiencies of different types of screens.

SCREEN/FILM
INTERACTIONS

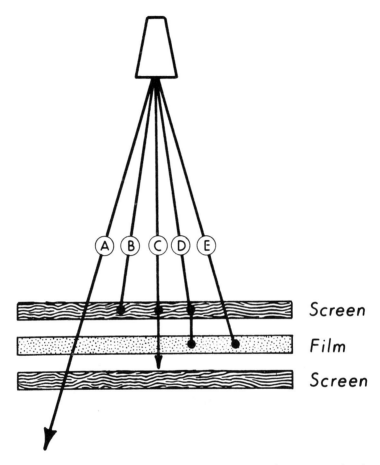

Figure 3-8. X-ray intensifying screen efficiency. An x-ray photon emanating from the x-ray tube has the probability of passing through both screen and film with no interaction **(A)**; being absorbed in the screen **(B)**; producing in the screen a light photon which does not contribute to the production of the latent image **(C)**; producing a light photon which does contribute to the latent image **(D)**; or producing a latent image in the x-ray film itself **(E)**. The efficiency of the intensifying screen is therefore very low, ranging from 2 to 20 percent for calcium tungstate screens and around 40 to 60 percent for rare earth screens.

Screen speed

The more efficient a screen, the less the radiation required to produce a certain amount of film density. There is, however, a point at which a uniform blackening does not occur simply because there are not enough photons striking the phos-

Table 3-1. Efficiency of intensifying screens

Screen	Percent of x-ray beam absorbed by screens
Detail	2–5
Medium-speed	20
High-speed	40
Rare earth	40–60

The efficiency of intensifying screens as demonstrated by the percentage of the diagnostic x-ray beam that is absorbed by a pair of screens. These numbers are estimates only, since screens will vary in speed from one manufacturer to another. The conversion efficiency—x-ray photons converted to light photons—is about 5 percent for calcium tungstate and 20 percent for rare earth phosphors. The efficiency of the film using the light photons emitted from the screens varies from 50 to 100 percent.

phor. In order to have uniform film blackening, the photons coming from the x-ray tube must be distributed uniformly across the film. It is well known that x-ray photons do not come from the x-ray tube in a uniform manner, but rather are distributed at random.

As an analogy to this problem, consider how raindrops fall on a dry surface. If one were to draw off a number of squares on a piece of dry concrete and then notice how raindrops fall on these squares, one would note that the raindrops do not cover each square in a uniform manner. However, after a short period of time, one notes that each square is uniformly covered. But suppose the rain stopped just when each square was barely wet. One would then note that even though each square looks wet, in fact within each square small areas are wetter than others.

The same effect occurs with highly efficient intensifying screens, with which relatively small amounts of radiation are required to produce what looks like uniform film blackening. On close examination, some small areas of the film prove to have more density than others. This effect is produced by the random distribution of x-ray photons striking the screen. The effect is called *quantum mottle.*

To tie things together, the grain or mottle one sees on a radiograph is composed of film granularity, screen structure mottle, and quantum mottle, collectively called *mottle* or *grain.* The major component of radiographic mottle is quantum mottle, not only with rare earth type screens but also with calcium tungstate screens. Screen structure mottle is a negligible factor; film granularity is a minor but not a negligible factor. With high-speed films, film granularity increases, but quantum mottle greatly increases.

Screen speed for calcium tungstate systems is achieved by varying the thickness of the phosphor layer, and to a much lesser extent by increasing the size of the phosphor crystals. While many textbooks relate screen speed to the size of the crystals, it is not a meaningful factor. The thicker the phosphor layer, the faster is the screen. Unfortunately, for calcium tungstate screens the thicker

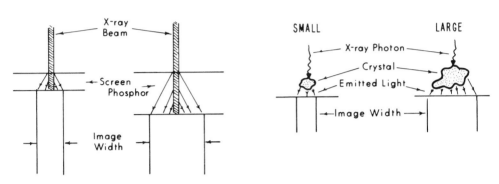

Figure 3-9. The effect of screen thickness and crystal size on radiographic details. Thick screens spread the image more than do thin screens. Large crystals also spread the image, but the contributions of large crystals are often overrated. Crystal sizes range from 4 microns in slow screens to 8 microns in very fast screens.

the phosphor, the greater the spreading of the radiographic image; and thus loss of detail becomes important. As shown in Figure 3-9, what happens in thick screens is that if a light photon is emitted from the anterior portion of the screen at a small angle, the farther this photon has to travel to strike the film, the farther from the center of the image it goes. This transverse travel (lateral spread) becomes worse as the screen phosphor is thickened. The result is that as the screen phosphor is thickened, greater degradation of the image occurs. Calcium tungstate screens have been improved to a point where there is an optimum marriage between resolution and speed (Figure 3–10).

At the present time, there are four general classes of screens, based largely on application and speed. These speeds are detail speed, medium or universal speed, high-speed, and rare earth. (The term *Par* is applied to medium-speed screens, but this is a trademark of Du Pont's medium-speed screens. The term *rare earth* should not be used to denote screen speed, but is used here for want of a standardized term. There are rare earth film/screen combinations which are slower than a medium-speed calcium tungstate film/screen system.) The most widely used screen is the medium-speed calcium tungstate; high-speed calcium tungstate screens are a close second internationally. In the United States, high-speed calcium tungstate screens are by far the most commonly used. Each manufacturer has his own trade name for screen speed, and it has been reported from an evaluation of medium or universal screens that speed from one manufacturer to another varied as much as 300 percent. This creates serious problems in an x-ray department which is not standardized on one brand of screens. It follows that, if a department uses a mixture of screens manufactured by different companies all of which are labeled universal or medium-speed screens, proper x-ray techniques cannot be formulated unless the screen speed of each is known.

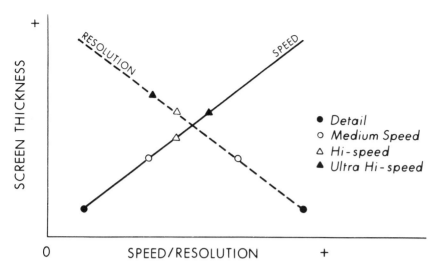

Figure 3-10. Speed and resolution of calcium tungstate intensifying screens. As the thickness of the screen phosphor increases, the speed of the screen increases; however, the resolving power of the screen correspondingly decreases.

DETERMINATION OF SCREEN SPEED

Now that so many new intensifying screens are available from manufacturers, it becomes important to develop a method for comparing one particular phosphor with another in speed. A simple method is to use the inverse square law. As described more fully in Chapter 7, the inverse square law states that radiation intensity is inversely proportional to the square of the distance from the source to the image receptor. Thus, if an exposure is made at 100 cm and it is desired to decrease the source/image-receptor distance to 50 cm, an exposure of one-fourth of the original would be required. This also tells us that if distance is increased, film blackening will be inversely decreased, and this knowledge can be used to determine screen speed. To determine screen speed using the inverse square law method, follow these steps:

1. Select a step on the step wedge which is 2–3 cm thick. On most step wedges, this is step 6 or 7. Selection of this step as our standard allows comparison of speed to be made at just about the speed point which is measured at 1 density unit above base-plus-fog density.
2. Expose the wedge on the current system. Exposure factors of 75 kV, a 30-in. source/image-receptor distance and 100 ms are suitable. Sufficient mAs should be used to produce a net density of 1.2 to 1.5 on the selected step.
3. Expose the test film/screen combination using the identical exposure factors

used in step 2. (The same cassette should be used, if possible. Needless to say, the same exposure and processing equipment should be used also.)

4. Compare the densities between the current and the test systems at the selected standard step on the step-wedge exposure.

5. Repeat the exposure, raising or lowering the SID (source/image-receptor distance) to match the density on the selected step of the step wedge.

In many cases where the speed of the new system is much higher than the current system, the film may be so dark that multiple exposures will have to be made before the densities come close together. After the first exposure, an estimate can be made as to what the next distance should be. Each step on a $\sqrt{2}$ step wedge requires about 6 in. in distance change to match densities. Thus, if step 3 of the test system is close to the density of our standard (say step 7), the next exposure should be made with the SID increased 24 in. (4 steps $\times$ 6 in.). After the second exposure, minor changes can be made in distances to exactly match the density of the test and current systems.

6. When densities on the selected steps are matched, then refer to Table 3–2 to determine the speed factor. In the example above, if the initial exposure was made at a 30-in. SID and it required 61 in. to match densities, in Table 3–2 we note that the speed of the test system is 4.13 times as fast as the current system at 75 kV. Speed changes may not necessarily be the same at all kV levels.

Table 3-2. Speed determination, using the inverse square law

Inches	Speed Factor	Inches	Speed Factor	Inches	Speed Factor
25	0.69	46	2.35	66	4.84
26	0.75	47	2.45	67	4.99
27	0.81	48	2.56	68	5.14
28	0.87	49	2.67	69	5.29
29	0.93	50	2.78	70	5.44
30	1.00	51	2.89	71	5.60
31	1.07	52	3.00	72	5.76
32	1.14	53	3.12	73	5.92
33	1.21	54	3.24	74	6.08
34	1.28	55	3.36	75	6.25
35	1.36	56	3.48	76	6.42
36	1.44	57	3.61	77	6.59
37	1.52	58	3.74	78	6.76
38	1.60	59	3.87	79	6.93
39	1.69	60	4.00	80	7.11
40	1.78	61	4.13	81	7.29
41	1.87	62	4.27	82	7.47
42	1.96	63	4.41	83	7.65
43	2.05	64	4.55	84	7.84
44	2.15	65	4.69	85	8.03
45	2.25			86	8.22

Film/screen combinations

Though the standardization of screen speed presents a major problem, film/screen mismatches do not seem to cause much difficulty (consult Table 3–3). Still it is important for beginning technologists to have an understanding of common film/screen combinations. In general, the medium-speed screen with a medium-speed film provides the best overall balance between speed and resolution, but the most frequently found combination is high-speed screens and medium-speed film. Table 3-3 lists some common combinations.

Table 3-3. Opinions on certain film/screen combinations [a]

Anatomical area	Screen	Film	Product
Chest	Slow	Any	Non-applicable
	Medium	Slow	Unacceptable due to high radiation exposure
		Medium	Satisfactory with unnecessarily high radiation dose
		Fast	Unsatisfactory due to high mottle
	Fast	Slow	Satisfactory but with an increased dose level
		Medium	Satisfactory: preferred
		Fast	Satisfactory
	Ultra-fast speed (includes rare earth screens)	Slow	
		Medium	
			Not determined
		Fast	
		Ultra-fast	
Small extremities			
Fractures Bone alignment Hardware alignment Dysplasias	Ultra-fast	Medium	Preferred
Dystrophies	Fast	Medium	Satisfactory
Trabecular pattern		Fast	Satisfactory
Osteomyelitic Fatigue fracture	Fine or detail	Slow or medium	Preferred
Head and neck	Slow	Any	Unacceptable due to high radiation exposure
	Medium	Medium	Preferred
		Fast	Satisfactory
	Fast	Medium	Satisfactory
		Fast	Unacceptable due to high noise level
	Ultra-fast	Any	Undetermined

Table 3-3. cont.

Trunk	Slow	Slow	Unacceptable due to high radiation exposure
	Medium	Slow	Acceptable
		Fast	Acceptable to preferred
		Medium	Preferred
	Fast	Medium	Preferred to acceptable
		Fast	Unacceptable due to high noise level
	Ultra-fast	Medium	Acceptable
		Fast	Acceptable
		Ultra-fast	Unacceptable due to high noise level

[a] Abstracted from the Film Conference of the American College of Radiology and the Bureau of Radiological Health, Washington, D.C.

There are many situations in which high-speed screens are very valuable and the slight loss of detail or resolution of secondary importance. The most common of these are:

1. Where reduction of radiation dosage is important; e.g., pelvimetry;
2. In portable bedside and operating-room radiography where there is a limit on kilovoltage and milliamperage;
3. In grid cassettes, because of the need to increase the kilovoltage to penetrate the grid;
4. In high-kilovoltage radiography;
5. In the radiography of infants and children, where speed is important;
6. In spot-film radiography, to decrease motion unsharpness.

Slow-speed screens render better detail and are used for bone radiography and some magnification techniques where speed is not at a premium (Figure 3-11).

Multisectional screen books for use in body-section radiography represent a special application of intensifying screens. A multisectional screen book is composed of a varying number of sets of screens of different speeds placed in the screen holder in such a way that the slowest pair of screens is closest to the x-ray beam, and faster pairs are farther away. With the multisectional screen book, up to seven layers of a body part can be radiographed simultaneously. Each screen pair is balanced in speed to yield the same exposure density on each film. Radiographic detail afforded by this system is not ideal, and such systems are seldom used.

A gradient intensifying screen is advantageous for venograms, femoral arteriograms, leg-length measurements, and radiography of the entire spine. In a gradient screen, the screen thickness gradually diminishes, so that the screen resembles a long, thin wedge filter. These screens are available in 14 x 36-in. and 14 x

51-in. sizes. The screen is positioned so that the thick anatomical part (e.g., abdomen) is matched with the thick part of the screen and thin anatomical parts (e.g., the lower extremity) are matched with the thin part of the screen. The screen thickness, therefore, compensates for increased tissue thickness. As a result, we obtain a uniform film density across decreasing anatomical thicknesses. A screen of this type eliminates the need for wedge filters by employing different screen speeds to produce uniform densities of the body parts being radiographed.

Screen speed, sharpness (resolution) and noise relationships

Until the introduction of rare earth intensifying screens a few years ago, a central topic of conversation among radiologists was the loss of sharpness of

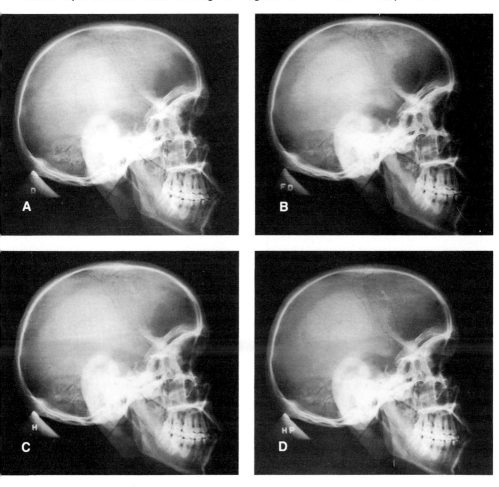

Figure 3-11. A. A radiograph made with a detail intensifying screen. **B.** A radiograph made with a fast detail intensifying screen. **C.** A radiograph made with a high-speed intensifying screen. **D.** A radiograph made with a high-plus intensifying screen.

an intensifying screen as its speed was increased. As noted above, this loss is due to the lateral dispersion of light as the photon travels through the screen. With rare earth screens, viewers have become more aware of the presence of *noise* (quantum mottle) in the radiograph. Nowadays the topic of conversation is speed versus resolution versus noise. With the higher-speed systems there is a trade-off, just as with standard-speed calcium tungstate intensifying screens; sharpness is traded off to reduce objectionable levels of noise. As an example, slow sharp screens are used to render better high-contrast bone detail. The use of slow screens with slow film allows us to greatly reduce the noise level for better visualization of soft-tissue details—e.g., in tomograms and gallbladder examinations. In the present state of the art, one cannot obtain maximum speed, least radiation exposure, highest resolution, and minimum noise. There is always a trade-off of one for another.

The response of screens to changes in kilovoltage

Intensifying screens do not respond linearly to changes in kilovoltage. The conversion efficiency remains unchanged, but the absorption varies according to kilovoltage. Throughout the years, technique charts have taken this fact into consideration in providing matched density exposures using high and low kilovoltage techniques. Many of the charts of the rare earth screens show the speed of the new screens relative to *normalized* speed of calcium tungstate screens. "Normalized" means that the speed of the referenced calcium tungstate system is assumed to be linear, and speeds of other systems are measured relative to the speed of calcium tungstate. The responses of some rare earth screens is more erratic to changes in kilovoltage than calcium tungstate screens. The reader should be cautioned that if rare earth screens are being used, he should check with the manufacturer's representative to ascertain within what ranges one can work with some assurance that technique factor changes will be appropriate. Use of rare earth screens dictates that accurate technique charts should be formulated and adhered to. Although technique guessing may sometimes work with medium- or slow-speed calcium tungstate screens, there is no room for guessing with some of the newer film/screen combinations. Figure 3-12 shows typical responses of intensifying screens to changes in kilovoltage.

Care of intensifying screens

Intensifying screens will last indefinitely if properly cared for. Dust and other foreign materials sometimes get into the cassette and will become embedded in the screen. Routine periodic cleaning will remove this material. The best cleaning agent is a mild soap and water. Detergents should not be used. Some technologists wipe the screens with 95 percent alcohol after washing with soap and water; the alcohol will evaporate rapidly and one does not have to wait long for the screens to dry. Alcohol may remove the top coat from some screens. It is best to check with the manufacturer as to how the screens should be cleaned.

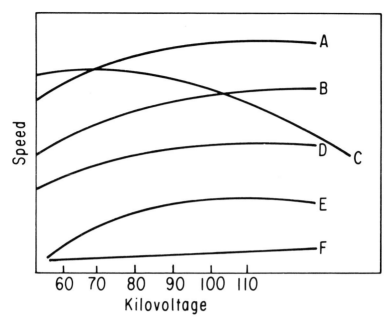

Figure 3-12. Responses of various intensifying screens to changes in kilovoltage. All screens are normalized to medium-speed calcium tungstate (not shown). Note that with the exception of curve F, which shows a slight increase in speed as kilovoltage is increased, the curves show varying degrees of speed changes and non-linear response to changes in kilovoltage.

If embedded material is not removed, the screen speed will decrease because of the foreign material blocking the light emitted from the screen. Screens with scratches in them must be discarded; scratches cannot be repaired. Similarly, small breaks will be made in screens if they are removed from a cassette, and these screens generally have to be discarded.

SUGGESTED READING

Ardran, G. M. "Testing X-ray Cassettes for Film-Intensifying Screen Contact." *Radiography* 35 (1969):143.

Balter, S. K. "The Incidence and Causes of Repeated Radiographic Examinations in a Community Hospital." *Radiology* 112 (1974):71.

Buchanan, R. A., et al. "X-ray Exposure Reduction Using Rare-Earth Oxysulfide Intensifying Screens." *Radiology* 10 (1972):185.

Buchanan, R. A., Finkelstein, S. I., and Wickersheim, K. A. "X-ray Exposure Reduction Using Rare-Earth Oxysulfide Intensifying Screens." *Radiology* 105 (Oct. 1972):185–90.

Crooks, H. E. "Some Aspects of Radiographic Quality with Special Reference to Image Sharpness." *Radiography* 39 (1973):317.

Gray, J. E. "Light Fog on Radiographic Films: How to Measure It Properly." *Radiology* 115 (1975):225.

Haus, A. G., and Rossman, K. "X-ray Sensitometer for Screen-film Combinations Used in Medical Radiology." *Radiology* 94 (1970):673.

Hess, I. I. "What We Should Know About X-ray Intensifying Screens." *Radiol. Tech.* 43 (1971):117.

Rao, G. U. V. "Do High Detail Screens Always Yield Better Resolution Than High Speed Screens?" *Amer. J. Roentg.* 112 (1971):812.

Reichmann, S., and Helander, C. G. "Homogeneity of Intensifying Screens." *Acta Radiol. (Diag.)* 15 (1974):449.

Split-Second Exposure (Intensifying Screens). Du Pont, Wilmington, Del. (No date.)

Stevels, A. L. V. "New Phosphors for X-ray Screens." *Medicamundi* 20, no. 1 (1975):12–22.

Thompson, T. T. "Selecting Medical X-ray Film—Part I." *Applied Radiology* 3 (Sept./Oct. 1974):47–50.

Thompson, T. T. "Selecting Medical X-ray Film—Part II." *Applied Radiology* 3 (Nov./Dec. 1974):51–53.

REVIEW QUESTIONS

Choose the correct response to the questions below.

1. A warped cassette results in

    a. Increased film density
    b. Blurring of the radiographic image
    c. Increased absorption by the cassette
    d. Decreased x-ray exposure to the patient
    e. No effect on finished radiograph

2. Screen contact is tested for by

    a. Making an exposure of a cassette
    b. Making an exposure of the screens and cassettes
    c. Placing a step wedge on the cassette and making an exposure of screen, cassette, and step wedge
    d. Making an exposure of a fine wire mesh placed on top of a cassette and screen
    e. There is no effective way for testing for screen contact

3. A common cause of transitory poor screen contact is

    a. A warped cassette front
    b. Air trapping within the cassette
    c. A warped grid
    d. Using the wrong type of film in the cassette
    e. None of the above

4. The light output from calcium tungstate screens is in what light spectrum?

    a. Red
    b. Blue
    c. Green
    d. Yellow
    e. Purple

5. Poor screen contact results in

    a. Half-moon shadows on the film

    b. Black streaks on the film

    c. Blurring of the radiographic image

    d. Decreased time in the developer

6. The back of a cassette is made of material opaque to x-rays because

    a. It tends to protect the technologist from scattered radiation

    b. It tends to reduce distortion in the radiograph

    c. It tends to prevent fogging of the film by scattered radiation from behind the cassette

    d. It further intensifies the image

    e. It tends to reduce the amount of exposure necessary

7. To reduce exposure of the patient to x-rays, the best screen to use is

    a. Non-screen film

    b. Par-speed screen with regular film

    c. High-speed screen with high-speed film

    d. Slow-speed screen with high-speed film

8. Poor screen contact will result in

    a. Increased sharpness of detail in the radiograph

    b. Increased contrast in the radiograph

    c. Decreased sharpness of detail in the radiograph

    d. Increased density in the radiograph

    e. Screen lag

9. The sharpness of detail of a thin part is increased by the use of

    a. Non-screen exposure

    b. High-speed screens

    c. High-speed film

    d. A grid

10. Screens should be replaced

    a. When speed decreases as compared with a department standard

    b. When artifacts are noted in the screen

    c. When brands of film are changed

    d. A, B, and C are correct

    e. A and B are correct

11. Screens are cleaned

    a. With a mild soap such as Ivory soap

    b. With scouring powder

    c. With household detergents

    d. With fine steel wool

    e. According to the manufacturer's specifications

12. The phosphor most commonly used in manufacturing intensifying screens is

    a. Silver iodide

    b. Silver bromide

    c. Calcium carbonate

    d. Calcium tungstate

    e. Yttrium oxysulfide

13. Newer phosphors used in manufacturing intensifying screens include all of the below EXCEPT

    a. Gadolinium oxysulfide

      b. Lanthanum oxybromide
      c. Yttrium oxysulfide
      d. Cesium iodide
      e. Barium fluorochloride

14. Increased speed of calcium tungstate screens is obtained primarily by

      a. Using small crystals
      b. Using no crystals
      c. Decreasing the thickness of the phosphor
      d. Increasing the thickness of the phosphor
      e. Addition of a dye to the binder

15. The primary light output from a gadolinium oxysulfide screen is in what light spectrum?

      a. Red
      b. Blue
      c. Green
      d. Yellow
      e. Purple

16. An x-ray intensifying screen is composed of all of the following EXCEPT

      a. A protective coat
      b. A phosphor layer
      c. A reflective coat
      d. A base
      e. An emulsion layer

17. Dyes are added to intensifying screens primarily to

      a. Increase the speed of the screen
      b. Change the color of the radiographic image
      c. Make it easier to detect defects in the screen
      d. Control lateral dispersion of light
      e. Control scattered radiation

18. High-speed calcium tungstate screens are approximately how much faster than medium speed calcium tungstate screens?

      a. 1×
      b. 2×
      c. 3×
      d. 4×
      e. 5×

19. Quantum mottle is due to

      a. Clumping of screen phosphor
      b. Statistical distribution of photons striking the phosphor
      c. Film granularity
      d. Improper film/screen contact
      e. Too much mAs

20. Mottle includes all of the below EXCEPT

      a. Screen structure mottle
      b. Film granularity
      c. Quantum mottle
      d. Screen contact

4 CONTROLLING SCATTERED RADIATION

Accessories such as cones, lead diaphragms, collimators, and grids have one thing in common: they control the scattered radiation striking the film. The beam-limiting devices (cones, lead diaphragms, and collimators) restrict the size of the x-ray beam (primary radiation) as it leaves the port of the x-ray tube. On the other hand, filters and grids provide absorption of x-ray photons. Filters accomplish this by absorbing low-energy x-ray photons at the x-ray tube port; grids absorb scattered or secondary radiation as it leaves the patient and before the photons strike the image receptor.

When x-ray photons strike an object, a certain amount of secondary and scattered radiation is emitted in all directions. (Although scattered radiation and secondary radiation are physically different, here we will use "scatter" to include both.) The contribution of scatter to film density ranges from 10 to 90 percent depending on the beam quality and the size and density of the object being irradiated. Remnant radiation is that portion of the incident x-ray beam which emerges from an absorber without change in direction. Since scattered radiation is emitted in all directions, and thus strikes the image receptor from all directions rather than straight on, as with remnant radiation (Figure 4-1), there is blurring of the true image with a resultant loss of detail. Secondary scatter must be controlled if radiographic detail and visibility of that detail are to be optimized.

Because beam-limiting devices restrict the amount of radiation striking the object, less scattered radiation is produced when they are used, and therefore less scattered radiation reaches the film. The result is that the finished radiograph has more apparent detail because of higher contrast than a comparable one for which the beam has not been limited (Figure 4-2). The limiting or restriction of the x-ray beam with the use of a beam-limiting device is called *collimation.* "To collimate," in the terminology of radiography, means to restrict the x-ray beam to an area of anatomical interest. Hence, when one says that a film was not collimated, one means that the x-ray beam was not restricted to the field of interest and that there are areas of film included within the field of radiation which should not have been.

BEAM-LIMITING DEVICES

Beam-limiting devices include cones, lead diaphragms, and collimators. Each of these devices has specific advantages and disadvantages.

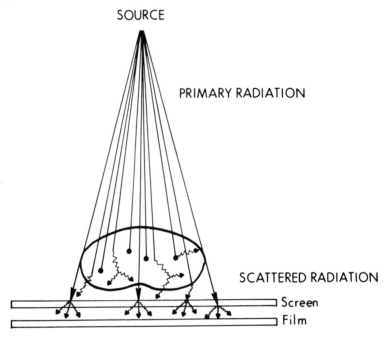

SOURCE

PRIMARY RADIATION

SCATTERED RADIATION

Screen

Film

Figure 4-1. The production of scattered radiation. X-ray photons leaving the x-ray tube interact with the body part and produce secondary and scattered radiation. Some of this radiation produces light photons in the screen. The light photons in turn produce a level of film blackening which is non-informational fog.

Cones

A cone is a cylindrical or conical tubular accessory which is attached to the x-ray tube port and serves the function of limiting the field of radiation emanating from the x-ray tube. With few exceptions, cones have been replaced by collimators. There are three major drawbacks to the use of cones in radiography. (1) A different cone is required for each different field size and for every SID (source/image-receptor distance). Since a complete set of cones would be required for each radiographic room, it is not economical nor practical to use such devices, except in special circumstances. (2) It is impossible for the circular cross section of the x-ray beam to be fitted to a rectangular film (Figure 4-3). The corners of the full-size film will be unexposed and the tissues adjacent to the edges of the film will be needlessly exposed to radiation. (3) With cones it is impossible to correctly align the x-ray beam with the anatomical area of interest; since the cone does not have a centering light, alignment is done by guessing at best. This often leads to misalignment of the x-ray beam with the film receptor.

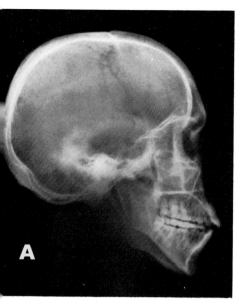

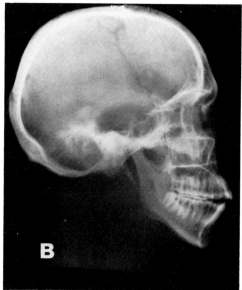

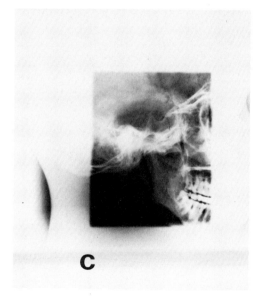

Figure 4-2. The effect of collimation on radiographic detail. In **A,** the collimator was opened as wide as possible; in **B,** the x-ray beam was collimated to the size of the film; in **C,** the beam was collimated to the region of the sella turcica. The film density in **A** is greater than for the other two exposures, and film detail increases as the x-ray beam becomes more restricted: 70-kV exposure, non-grid, using medium-speed calcium-tungstate screens.

Lead diaphragms or apertures

The lead diaphragm is a sheet of lead or lead plate which is attached directly to the x-ray tube port. A hole in the plate shapes the x-ray beam. The main disadvantage of the diaphragm is that it does not provide a sharp demarcation of the edge of the x-ray beam (Figure 4-4). Only rarely, because of their disadvantages, are lead diaphragms used in radiography. For the same reasons that make cones impractical, lead diaphragms are used only in special procedures, seldom

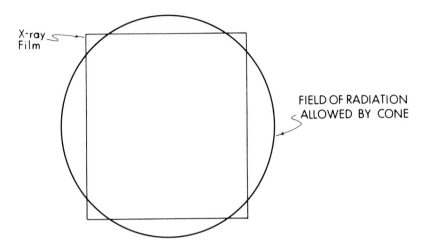

Figure 4-3. An example of cone-cutting. In this case the corners of the film are not exposed, and useless radiation exposure is allowed beyond the edges of the film.

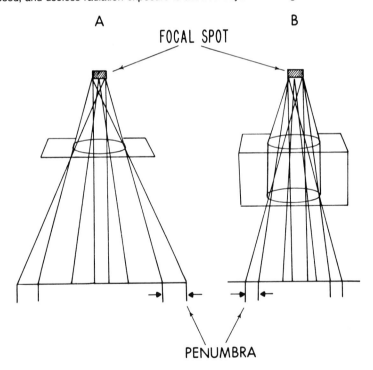

Figure 4-4. The effect on edge penumbra of the location of a cone or diaphragm. With a diaphragm **A,** off-focus is not controlled, and there is a large area of penumbra at the *edge* of the exposed film area. With a cylinder cone **B,** off-focus radiation is controlled much better, and there is a considerable decrease in edge penumbra. With well-designed collimators, edge penumbra is tightly controlled.

in general radiography. To use them invariably, means one has to remove the collimator, and then one is faced with the problem of realigning the collimator centering light to the radiation field. Some collimators have built-in slots which will accept diaphragms.

Collimators

Collimators—devices with adjustable lead shutters—are used to limit the x-ray beam and to center the beam over a specific anatomical area. Like cones and aperture diaphragms, they are attached directly to the x-ray tube port. The collimator (Figure 4-5) consists of a series of vertical and horizontal adjustable lead plates which slide in and out of the x-ray beam, thus limiting the cross-sectional area of the beam. Collimators also contain a built-in light source which assists the technologist in centering the x-ray beam over the field of interest. This light is located on the side of the collimator housing and is positioned in such a way that its light, reflected by a mirror, corresponds directly with the field coverage of the x-ray beam.

Correct alignment of the collimator centering light with the x-ray beam is ex-

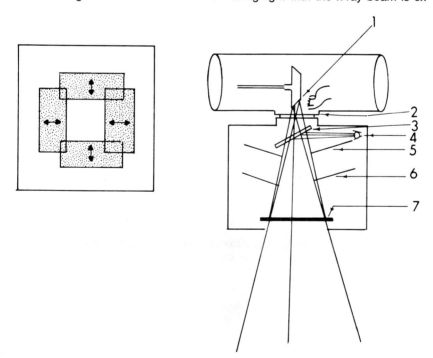

Figure 4-5. A modern collimator. **1.** Focal spot. **2.** X-ray tube port. **3.** Mirror. **4.** Light source. **5.** Upper shutters. **6.** Lower shutters. **7.** Added filtration. The schema on the left shows the movement of the shutters. The shutters can be electrically controlled so that they cannot be opened wider than the size of the image receptor. Notice how the double shutter system effectively controls edge penumbra.

tremely important, and all technologists should understand and be able to check the accuracy of the centering light. The centering light/x-ray beam alignment can be verified by simply taking a 35 × 42-cm (14 × 17 in.) cassette and loading it with film. Place the cassette on the radiographic table or in the Bucky tray and adjust the centering light so that all of the cassette is covered by the light field except for the peripheral 5 cm. Place a coin or other small metallic object at each corner of the light field. Make a 50-kV, 10-mAs exposure and process the film. The radiograph should show the coins to be no more than 2 cm inside the field of radiation. If they are more, the collimator centering light should be realigned by someone who has been trained to realign it. The adjustable collimator allows exact positioning and limiting of the beam area, so that the patient need not be exposed unnecessarily to radiation. Available beam shapes vary from regular polygons, which may be enlarged or made smaller, to rectangular slots with choice of axis. Figure 4-6 makes note of this.

The collimator takes the guesswork out of centering the part over the film. The collimator, correctly adjusted, shows exactly the body area to be exposed before the exposure is made. Scattered radiation is controlled. The result is better radiographic contrast, and a reduction in the exposure of the patient to radiation.

Automatic collimators

A major cause of overexposure of patients to radiation is improper collimation. It is not unusual to see radiographic exposures made with collimators wide open, even by well-trained technologists who should know better. By not coning to the field of interest, not only is tissue not in the field of interest given totally

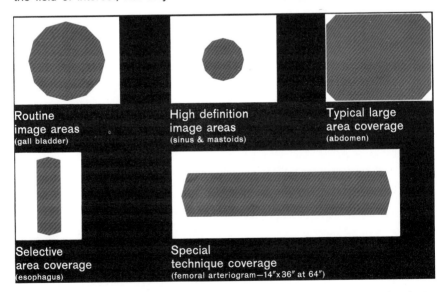

Figure 4-6. Some examples of beam shapes ("geometry") available from the shutter systems of a modern collimator.

useless radiation, but the open collimation allows more scatter, which contributes to film fog that degrades the radiographic image. In many institutions radiologists and chief technologists require that all films show cone cuts, to indicate that the x-ray beam has been collimated at least to the size of the film being used. A film shows a "cone cut" when it shows a small portion of its edges unexposed because of a slight undercollimating of the x-ray beam.

By federal regulation all x-ray equipment manufactured after August 1974 is required to have automatic collimation. Collimators are so manufactured and electrically interconnected to the Porter-Bucky diaphragm that the collimator automatically adjusts to the size of the x-ray film cassette placed in the Bucky tray. This automatic equipment is supposed to preclude unnecessary exposure of tissue which is not being recorded by the film. Unfortunately, most automatic collimators have manual override switches which allow the automation to be overridden and the collimator used like any other collimator; this manual override is often misused by technologists.

Filters

Photon energy is measured in electron volts (Ev). When, for example, 100 kV is applied to an x-ray tube, the maximum energy of any one x-ray photon is 100 kiloelectron volts (kEv), because the maximum amount of energy that can be transferred from an electron striking an anode is the maximum amount of energy carried by the electron. If, for example, 100 kV is applied to the x-ray tube, the maximum energy of the electron striking the anode is 100 kEv. Not all electrons produce a maximum energy transfer; thus, a variety of photons with wavelengths encompassing heat, light, and x-ray spectrums are emitted in the form of a photon flux or spectrum. Since an x-ray beam is composed of a flux of photons, most of the photons will be at an energy level less than 100 kEv. For a 100-kV exposure the average photon energy will be around 77 kEv. By hardening the beam, i.e., filtering out low-energy photons, the average energy of the remaining photon flux is made higher. The Department of Health, Education, and Welfare (HEW) regulations require that a minimum of 2.5 mm of aluminum filtration (total, or added to inherent) be placed in the x-ray beam. This aluminum provides absorption of long-wavelength, low-energy photons which contribute significantly to patient exposure but almost nothing to the radiographic image. In most cases 1.5 mm of this filtration is provided by the window of the x-ray tube, and 1.0 mm by the collimator (see Figure 4-5). In addition to this filtration, some collimators are provided with slots for additional filters. Others have rotating filter discs which can be used to add little or much filtration to the beam. As more filtration is added, more of the soft x-ray photons are absorbed, with the result that the x-ray beam is hardened, i.e., the average photon energy (kEv) of a quantum of photons is increased. In the diagnostic range of kilovoltages the additional filtration required by HEW produces no significant visible effect on the radiograph. As more filtration is added, there is a concomitant increase in latitude, such as might be expected if one used higher-kilovoltage techniques.

Figure 4-7. A wedge filter is shown on the left and a trough filter on the right. The thicker part of the filter allows less radiation to pass through it and thus less film density results. The thick part is arranged so that it will hold back radiation to the anatomical part that is generally overexposed, while the thin part allows an underexposed anatomical area to catch up.

Wedge and trough filters

The wedge and trough filters have special applications, specifically in chest radiography and angiography and occasionally in radiography of the extremities. Figure 4-7 shows examples of these filters. Wedge and trough filters provide a uniform density across the radiograph where body parts vary widely in opacity. For example, it is extremely difficult to expose a chest radiograph so that there is sufficient density and penetration of both the mediastinum and the periphery of the lung fields. If the mediastinum is adequately penetrated and has an acceptable density on the film, the periphery of the lung fields is overexposed, and vice versa. A wedge filter can allow more radiation to penetrate the hilum of the lung and restrict the amount of radiation striking the periphery of the lung. As a result, a single exposure can be made which will allow one to see both the periphery of the lung and hilar areas at the same time. Thus, one film is used in lieu of two. A trough filter can be used when both lungs need to be seen along with the mediastinum. The main disadvantage of most wedge and trough filters is that they have to be placed in or on the collimator after the x-ray beam is centered. After the filter is in place, the collimator light is blocked and therefore not available to the technologist. Wedge filters can be useful in radiographing extremities, again for the same reason; if the thin part of the extremity, for example a phalanx, is adequately exposed, the carpal bones of the wrist are generally underexposed.

THE EFFECTS OF COLLIMATION

By collimating, the quality of the film is directly affected in two major ways: (1) exposure of the film is decreased, and this is usually associated with a net overall decrease of film density, and (2) visibility of detail is increased.

As already noted, scattered radiation contributes more to film density than does remnant radiation. If the incident x-ray beam remains constant and by some mechanism scattered radiation is reduced, net film density has to be reduced. The reason is this: the greater the volume and mass of material there is to be radiated, the greater is the amount of scattered radiation (within the limits of beam quality and density of material). Thus, in irradiating human tissue, the larger the area irradiated, the more scatter there is produced. The scatter lays down a veil of density across the irradiated film, but adds nothing to contrast or visibility of detail. This veil of density is sometimes called "non-informational density." When the x-ray beam is collimated (Figure 4-8), a smaller amount of tissue is irradiated, less scatter reaches the film, and the result is an overall decrease in film density. It follows that the more collimation there is, the more one must compensate by increasing kV or mAs or both so as to achieve an adequate film density. Increased film density can be obtained by using other techniques, to be discussed in Chapter 5.

With collimation, the veil of density which clouds detail is partially removed from the film, so that one can better visualize low-level contrast; density difference between low-level densities is better appreciated, and therefore one can see

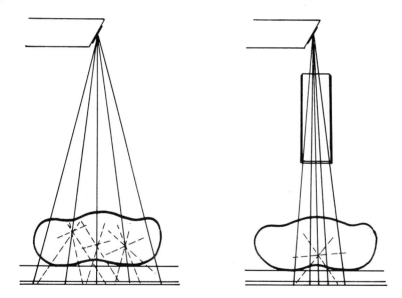

Figure 4-8. The effect of the cone upon density and contrast. In these diagrams, it is to be assumed that only the central portion of the body section shown is of diagnostic interest. In the first diagram, without cone, it will be seen that secondary radiation set up in portions of the body outside the area of interest but within the area irradiated by the x-ray tube can reach the film; the resulting secondary-radiation fog will increase density and reduce contrast in the radiograph. The second diagram demonstrates the use of a cone to restrict radiation to the area of interest. Since the area in which the secondary radiation can be set up is reduced, secondary-radiation fog density in the resulting radiograph will be reduced and contrast will be improved.

small detail better. The removal of the veil thus allows greater visibility of detail.

There is often a misunderstanding concerning the effect of collimation on contrast, density, and visibility of detail. To more fully explain the effect of scattered radiation, we need to define terms as accurately as possible. *Contrast* by definition means density difference. Contrast can be determined from the straight-line portion of a sensitometric curve by calculating average gradient as explained in Chapter 2 and Figure 2-17. *Density* is defined by the percentage of light transmitted and is expressed in log units. In order to have contrast, one must also have density. *Sharpness* is a function of a number of items—focal spot size, source/object and source/image-receptor distance, physical parameters of the recording system, and motion of the object being radiographed. In order to see sharpness, one must have contrast, and in order to have contrast, density must be available. Sharpness of detail is related to sharpness and is measured by use of edge gradients. An edge gradient is determined by physically measuring how abruptly one level of density changes to another. The more abrupt the transition, the better the sharpness. Visibility of detail refers to how well one can visualize the detail already recorded on film. Visibility of detail is also related to the quality of the viewboxes, the hue and intensity of the light transmitted through the viewboxes, the subjectivity of the viewer, and a host of other things. When a film is fogged, whether from secondary radiation or any other cause, this fog masks or obliterates detail so that it is more difficult to see. By removing fog of any type, detail can better be visualized.

As far as scattered radiation is concerned, a non-informational fog level is produced all across the exposed area of the film. The more the scatter, the higher the fog level will be. Very low levels of fog affect the toe of the sensitometric curve and create no significant effect on the straight-line portion of the sensitometric curve. Hence, low-density detail is obscured by secondary fog, but film contrast as measured by sensitometry is not. As the fog level increases, more and more of the curve is affected, and contrast is more and more affected. (Figure 4-9 illustrates this more clearly.) Therefore, contrast-density difference between sensitometric steps decreases as fog increases. The effect of collimation is to partially remove this veil of fog, allowing one to see what has already been recorded—that is, to allow an increase in visibility of detail.

FACTORS AFFECTING SCATTER RADIATION

Two main factors affect scattered radiation—density and thickness of the body tissue being irradiated, and selected kilovoltage. The thicker the tissue, the more scattered radiation there is. The contribution of fog to film density is greatest from the portion of the body part that is farthest away from the film and least from the portion of the body part that is closest to the film. The reason for this is that the closer the body part is to the film, the less distance the photons have in which to cross over or undercut the body part and create spurious images on the film. Dense structures with high atomic numbers, such as bone, absorb

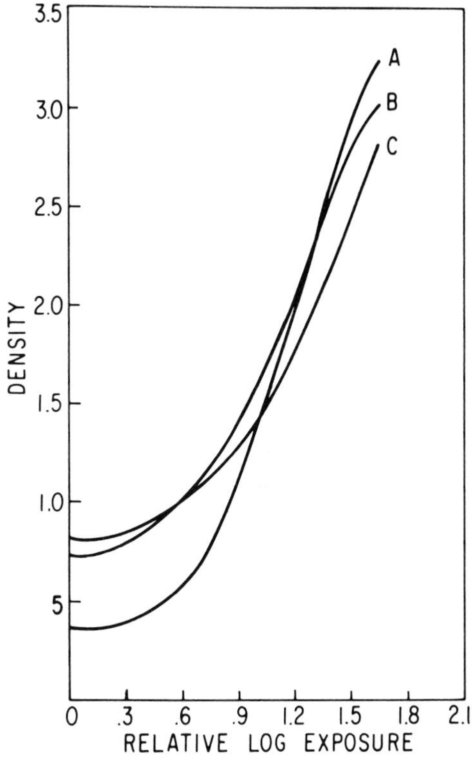

Figure 4-9. The effect of scattered radiation on sensitometric curves. Curves were plotted from step-wedge exposures, with the step wedge placed 5 cm above a cassette containing medium-speed screens and medium-speed film. Curve **A** is an exposure of a film without absorber; curve **B**, of a film with a 2-cm Plexiglas absorber; curve **C**, of a film with a 5-cm absorber. Curve **A** shows low base-plus-fog level and medium contrast; curve **B** shows an increase in base-plus-fog level, plus a mild loss of contrast; curve **C** shows a further increase in fog level and a further loss of contrast. The changes in the sensitometric curve are entirely due to the addition of scattered radiation to the film.

more radiation and therefore produce less scatter. Dense structures with low atomic weight, such as muscle, produce more scattered radiation than those with higher atomic numbers.

The higher the kilovoltage, the more scattered radiation there is. As the peak or highest energy in the photon flux increases, e.g., by increasing the kilovoltage, not only does the peak kilovoltage increase but the average energy of individual photons also increases. The higher the energy of the incident or incoming photon, the higher the amount of scattered radiation and the greater the amount of this radiation that is scattered forward; and thus, the more scatter there is available to produce fog on the film. Thus, by increasing kilovoltage, as may be required to reduce radiation exposure to the patient or to change contrast, the more secondary fog there is produced on the film. That is why a technologist must learn optimal kilovoltage levels for all radiographic examinations. This optimum kilovoltage must take into consideration the contrast level required for the examination to be performed (the primary concern) and the amount of scattering produced by the kilovoltage level (the secondary concern).

Control of secondary radiation by beam-limiting devices and grids is an important fundamental which must be understood by the student in order to produce quality radiographs. Secondary radiation can be controlled by the appropriate application of the beam-limiting device, the proper selection of kilovoltage, and the proper use of grids.

RADIOGRAPHIC GRIDS

The grid diaphragm was invented by Gustav Bucky of Germany in 1913. At first it was not very practical, since it was used in a stationary position. Dr. Hollis Potter of Chicago made the Bucky grid practical by moving it during the radiographic exposure, a procedure that blurred the grid lines out of the image. The grid has since been known as the Potter-Bucky diaphragm, which is also the correct NEMA term for the device. In addition to proper collimation, the most effective way of removing scattered radiation from the radiograph is with grids. However, no grid can compensate for improper exposure techniques and improper collimation.

A grid is a device placed between the patient and the image receptor which serves the purpose of absorbing scattered radiation. Normally, the grid is attached to or part of the film cassette or is located in a Potter-Bucky device. Grids are constructed of strips of lead which alternate with strips of radiolucent material such as fiber, wood, plastic (organic), or aluminum (inorganic) (Figure 4-10). The lead strips are oriented in relation to the x-ray beam in such a manner that most of the primary beam passes between the strips of lead and most scattered

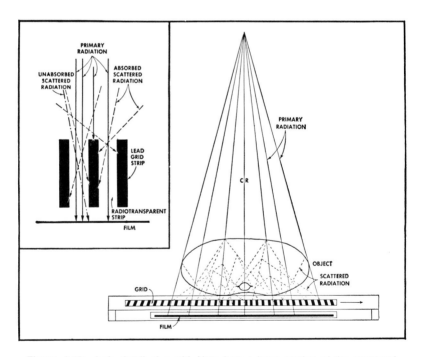

Figure 4-10. Left, detail of a grid. Note how a large portion of the scattered radiation is absorbed, but image-forming radiation passes through. Right, diagram of a Potter-Bucky diaphragm.

radiation will strike the strips and thus be absorbed. Since a grid will absorb a small portion of the primary beam and most of the scattered radiation, film density is decreased, and one must compensate for this by increasing the amount of incident radiation. The amount of increase will depend on the type of grid and a number of other factors which will be discussed later in the chapter.

TYPES OF GRIDS

Grids can be classified as being linear or crossed, and focused or non-focused. In addition, they may be moving (reciprocating) or stationary. Stationary grids are usually used in reciprocating devices. Reciprocating grids are commonly referred to as "Buckys," and we will refer to them thus in this book, although in fact it was Potter who invented the moving grid. Grids are further described in terms of grid ratio and lines per inch (strip density).

Linear and crossed grids

In a linear grid the lead strips are parallel to each other and will stop the photons which cross the axis but not those which are angled to the axis. A crossed grid, also known as a cross-hatched grid, is essentially two linear grids, one on top of the other, with the lead strips of one at right angles to those of the other (Figure 4-11). The crossed grid combines the ability of two linear grids to absorb radiation, since the scattered photons which can go through one grid by being angled to its axis will be absorbed by the other grid. Linear focused grids are the most commonly used in radiography. Less critical alignment of the x-ray beam with the grid is acceptable in the linear grid, whereas the crossed grid is more difficult to center under the x-ray beam. For linear grids the x-ray beam must be aligned with the center of the long axis; angulation with the long axis is possible. For crossed grids the beam must be aligned with the center of the grid; no lateral or longitudinal angulation is permissible.

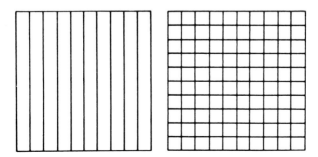

Figure 4-11. Two common grid patterns: linear (left) and crossed (right).

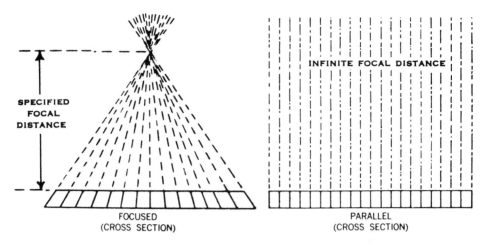

Figure 4-12. Cross sections of two grid arrangements: left, a focused grid with progressively angled leads; right, a parallel grid.

Focused and parallel grids

In focused grids, the lead strips are angled so that if a line is drawn from each strip, all of the lines will converge at a point in the midline and above the grid (Figure 4-12). The distance above the midline of the grid where these imaginary lines converge or intersect is known as the *focal distance* of the grid. Focal distances of grids are inscribed on the grid in a conspicuous place, and one should form the habit of checking the focal distance of a grid before it is used in exposing a radiograph. If an exposure is made within the prescribed focal distance of the grid, the radiation transmitted through the grid will be essentially uniform. If an exposure is made above or below this focal distance, grid cutoff will occur. (See the section on faulty use of grids, page 101 below.)

A parallel grid, also called a non-focused grid, is one in which the lead strips are parallel to each other and are focused to infinity (Figure 4-12). Theoretically, if lines were drawn from each lead strip, these lines would be parallel to each other and never converge to a point. Parallel grids are not often used in radiography. Although the photons in the center of the x-ray beam will pass through a parallel grid, the lower edges of the outer strips of the grid will provide some absorption of the x-ray beam. Unless a very long focal-film distance is used, the effect on the radiograph is a decreasing density towards the sides of the film, with normal density seen at the top and the bottom. So, in practice, parallel grids are used only for small-area radiography or for very long source/image-receptor distances.

Stationary and reciprocating grids

As its name implies, a stationary grid remains stationary during the time of the radiographic exposure. It is usually fixed temporarily by use of a holder or is permanently mounted on a film cassette. There are numerous applications where there is a need to eliminate scattered radiation, but where the reciprocating grid cannot be used. The stationary grid is used for fluoroscopy, portable radiography, spot-film radiography, and unusual techniques such as large chest radiography and cervical spine radiography. In some institutions, all exposures of the chest are made with stationary-grid techniques. The disadvantage of using stationary grids is that grid lines are usually visible and there is some difficulty in accurately aligning the x-ray beam with the grid.

Reciprocating grids or Buckys are moved during the time of exposure. Three possible motions are available; single-stroke (one-way), reciprocating, and catapult. The moving of the grid has the effect of eliminating grid lines. As the grid moves, the lead strips go in and out of focus. Reciprocating grids are usually mounted permanently in radiographic tables or on upright cassette-holding devices. As with stationary grids, an exposure must be made within the focal distance of the grid.

GRID RATIOS AND GRID LINES

By definition, grid ratio is the ratio of the height of the lead strips to the width of the interspace material between them. Notice that grid ratio does not take into account the *thickness* of the lead strip. Grid ratio can be used to indicate the efficiency of the grid in removing scattered radiation. Common grid ratios include 5 : 1, 6 : 1, 8 : 1, 10 : 1, 12 : 1, and 16 : 1. Crossed grids use the same terminology and indicate two linear grids with a given grid ratio. For example, an 8 : 1 crossed grid would indicate two 8 : 1 linear grids which are placed one on top of the other so that the strips are at right angles. The higher the grid ratio, the more efficiency or "cleanup" there is of scattered radiation (Figure 4-13). By the same token, the higher the grid ratio, the more the exposure tech-

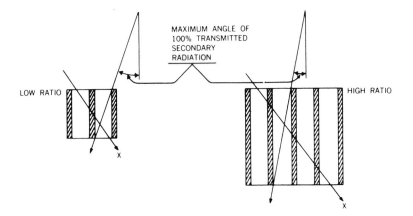

Figure 4-13. The effect of ratio on cleanup.

niques must be increased to obtain a satisfactory film density. In removing scatter radiation by using a grid, film contrast is also improved.

As indicated above, grid ratio does not take into consideration the thickness of the lead strip. This is accomplished by indicating the number of lead strips per unit width or lines per inch. Grids are available in 65 to 110 lines per inch. Measurement of grids in terms of lines per inch is also known as *strip density*. Grids with low strip densities are not very useful in absorbing scattered radiation; on the other hand, grids with high strip densities require substantially more radiation exposure to the patient and are more difficult to center the x-ray beam over. Moreover, grids with low strip densities have prominent grid lines and those with high strip densities are ineffective in cleaning up scattered radiation. Therefore, one must strike a compromise between the need to keep the exposure of the patient to radiation low and the need to have better detail and contrast in the radiograph. This is usually accomplished by choosing a midlevel (8 : 1) grid with strip density of 85 lines per inch.

SELECTION OF A GRID

The selection of a grid to be used for a particular radiograph will be primarily dependent on the following considerations:

1. The relative quantity of secondary radiation produced by the subject being radiographed;
2. The kilovoltage technique to be used; and
3. The capacity of x-ray generator.

The quantity of secondary radiation produced is dependent on the thickness and relative density of the body volume being radiographed. A non-grid exposure of the chest will consist of about 50 percent secondary radiation, while a non-grid exposure of the abdomen may consist of more than 90 percent secondary radiation. From this, it is apparent that for dense body sections the more effective remover of secondary radiation will provide the more striking improvement in the radiograph, and this suggests the use of a high-ratio grid or a crossed grid. The choice between these two grids depends on the ease of aligning the grid correctly with the x-ray tube, and whether a high- or low-kilovoltage technique is to be used. If there is any question of the possibility of proper centering and leveling, or if low kilovoltages are to be used, the linear grid of 5:1 ratio will present much greater advantages from the standpoint of positioning latitude and cleanup. For high-kilovoltage techniques, particularly if the grid can be accurately aligned or if it is mounted in a Bucky, greater advantages will be gained with the high-ratio grid. Crossed grids are not recommended for the techniques that require angling of the x-ray tube and therefore are not generally installed in Buckys. However, crossed grids can be used in spot-film devices where tube-angle techniques are not possible.

At kilovoltages of the order of 100 kV or more, comparable photographic effect

Table 4-1. Grid ratios and kilovoltage

	Upper kV limits	
Grid ratio	*Bone studies*	*Barium studies*
8 : 1	90	100
12 : 1	100	120
16 : 1	150	150
	Chest studies	
Grid ratio	*kV range to avoid excessive contrast*	
6 : 1	100–120	
8 : 1	140	
10 : 1	150	

requires lower mAs values than at low kilovoltages to keep the radiation dosage to the patient low. However, in order to maintain the same contrast range at the higher kilovoltage, it is necessary to use a grid of high ratio. The exposure factors are not the same for all grid ratios, and the increased kilovoltage required for a high-ratio grid may to some extent reduce the patient-dosage advantage gained by going to the high-kilovoltage techniques. In general, in spite of the higher exposure factors involved, the use of high kilovoltage and high-ratio grids will result in somewhat lower radiation dosage to the patient, particularly if high-speed screens are used.

It is little known that grid selection should be placed in three separate categories in radiography. The upper limits and area of best efficiency are set forth in Table 4-1. As the table shows, e.g., an 8 : 1 grid should not be used above 90 kV for bone studies or above 100 kV for barium studies.

All technologists must work within the limitations of the physical characteristics of the x-ray equipment at their disposal. While this may not be as important a consideration in the selection of a grid as some others, it is a factor to be considered. Most modern x-ray generators are designed to operate at a maximum of 125 or 150 kV. In using higher kilovoltage techniques, more scattered radiation is produced, and this scatter is more in the forward direction than at lower kilovoltage levels. Hence, with higher kilovoltage techniques, higher grid ratios are necessary to clean up the scattered radiation. As a rule of thumb, 8 : 1 or lower ratio grids should be used below 90 kV; 10 : 1 or higher ratio grids require a higher kilovoltage. Table 4-2 shows typical characteristics of grids.

In converting a non-grid technique to a grid technique, we must increase either the mAs or the kV because of the absorption of the grid. The student should note that exposure factors increase with increasing kV for all grids. At 110 kV, using a 12 : 1 ratio grid, approximately a 50 percent increase is required at the 110-kV level as compared with a 60-kV level. Reciprocating grids require a slightly greater increase in technique factors as compared with stationary grids. The increase in technique must be individualized to each piece of equipment, since part of this increase is related to the speed of movement of the reciprocating Bucky grid. A grid should be used for any part which is solid, or where there is

Table 4-2. Grid characteristics

Type of grid	Features and uses
4 : 1 ratio, linear (low dose)	Special grid for use in image intensification and spot-filming. It offers adequate cleanup for small coned-down areas, combined with relatively small patient dosage.
5 : 1 ratio, linear	Moderate cleanup. Extreme latitude in use. Use at lower kilovoltages (up to 80 kV) wherever wide latitude is desired. Very easy to use.
5 : 1 ratio, crossed	Very high cleanup, especially at lower kilovoltages. Extreme latitude in use. Use up to 100 kV wherever wide latitude and excellent cleanup are desired. Very easy to use. Not recommended for tilted-tube techniques.
6 : 1 ratio, linear	Moderate cleanup. Good positioning latitude. Easy to use.
8 : 1 ratio, linear	Better cleanup than 5 : 1 linear. Fair distance latitude. Little centering and leveling latitude. Use up to 100 kV where wide latitude is not required.
12 : 1 ratio, linear	Better cleanup than 8 : 1. Very little positioning latitude. Use for both low and high kilovoltage techniques (up to 110 kV or slightly higher). Extra care is required for proper alignment in use. Usually used in a fixed mount or Potter-Bucky diaphragm.
16 :1 ratio, linear	Very high cleanup. Practically no positioning latitude. Intended primarily for use above 100 kV in a Potter-Bucky diaphragm. Excellent for high-kilovoltage radiographs of thick body sections.
8 : 1 ratio, crossed	Extremely high cleanup. Superior to 16 : 1 ratio linear grid at kilovoltages up to 125. Positioning latitude equivalent to 8 : 1 ratio linear. Not usable for tilted-tube techniques.
10 : 1 ratio, 133 line, linear	No longer manufactured.

edema or accumulation of fluid. In general, a grid should be considered for any part thickness which is greater than 12 cm, except for a PA (postero-anterior) chest. (Conversion factors for grid techniques are further discussed in Chapter 7 and the factors are shown in Tables 7-7 and 7-8.)

A simple experiment can be done to determine conversion factors for grids (Figure 4-14).

1. Make an exposure with a radiation meter in position A (Figure 4-14) and record the reading.
2. Insert a grid, move the meter to position B, and make an exposure while increasing mAs until the meter reading in position B is the same as in position A.
3. Divide mAs in (2) by mAs in (1) to get the correction factor for the particular grid (ratio, lines/in.) that was inserted.

This procedure can be repeated many times with different ratios, lines/in. grids, kV, part thickness, etc.

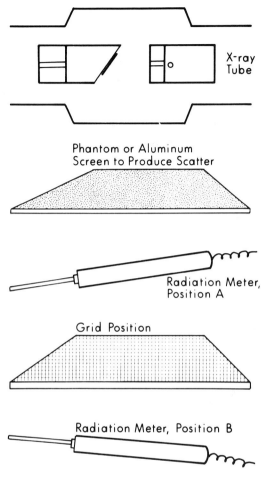

Figure 4-14. Determining conversion factors for grids.

FAULTY USE OF GRIDS

There are certain restrictions one must observe when employing a grid. These restrictions concern the *centering* of the x-ray tube, direction and degree of *tilt* of the x-ray beam, the choice of *source/image-receptor distance (SID)* and *selection* of kilovoltage.

Except for kilovoltage, all of these factors are involved with the direction of the primary beam in its passage through the grid. Ratio governs the amount of restriction placed on the technologist. A low-ratio grid is easy to work with because some off-centering and distance change can be tolerated, while a high-ratio grid requires very precise centering and operation only at the focal distance of the grid.

Grid cutoff

The major difficulty in the use of grids is grid cutoff, which occurs when the x-ray tube focal spot is not aligned correctly with the grid. As already noted, every grid is made to be used within a specific focal distance. If the x-ray tube is centered above or below, or to either side of this focal point, grid cutoff will occur. Grid cutoff is defined as the loss of primary radiation due to the improper centering of the x-ray beam over the grid. Most grids are manufactured so that there is some latitude in the aligning of the x-ray tube with the grid, and as long as one stays within this latitude, there will be no visible effect on the radiograph. High-ratio and cross grids require more care for proper alignment. Generally, short focal distance grids require more care in centering than long focal distance grids.

There are five major problems encountered in centering grids under the x-ray beam. These include:

1. Centering outside of tolerances of the designated focal distance of the grid;
2. Improper identification of the grid focal distance;
3. Lateral displacement of grid and/or tube;
4. Improper angulation of the tube and/or grid; and
5. Grid installed or used upside down.

Grids are labeled by the manufacturer as to grid ratio, strip density, and focal distance. Most x-ray departments have grids of several types, and it is a common mistake for a technologist to use one grid thinking it is another. For portable work, one should develop a habit of checking the grid every time one is used. At the time of checking to make sure one has selected the proper grid, the focal distance of the grid must be double-checked. The focal spot of the x-ray tube must be aligned with that prescribed focal distance. Centering outside of the prescribed focal distance (off-focus centering) will cause grid cutoff. Grid cutoff appears on the radiograph as an increasing loss of density towards the periphery of the film. Density along or parallel with the lead strips will be essentially uniform. Hence, film density will be maximum at the center of the film and will taper off towards the edge (Figure 4-15). All grids in permanent Bucky devices should be standardized.

In addition to placing the x-ray tube in the proper focal distance of the grid, it must also be aligned over the center of the grid. If lateral centering of the tube and/or grid (off-center) is not correct, none of the lead strips are aligned with the beam and there will be a uniform loss of density across the film. Errors from lateral misplacement of the grid under the x-ray beam are commonly mistaken for an incorrect exposure technique. Slight angulation of either the x-ray tube or grid will create the same effect. The x-ray beam or grid may be angulated as long as the angulation is with the axis of the grid and not across it. Thus, linear grids can be used for studies which require tilting of the tube or grid, provided the tilt is along the axis of the grid. Grids cannot be tilted across the

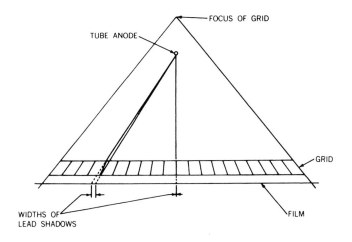

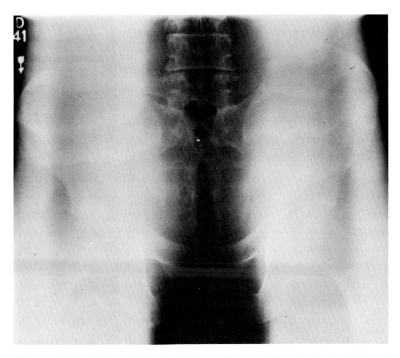

Figure 4-15. Grid cutoff due to off-focus alignment. The physical principle of off-focus grid cutoff is indicated in the diagram. In the radiograph, note the loss of density and detail on the right and left edges of the film, while proper density and detail are maintained in the center of the film. (Owing to change in focal-film distances and inverse square law restrictions, exposure techniques were varied to obtain matched densities in the center of the film. This is an exaggerated case.)

axis without loss of density nor can cross grids be tilted. Figure 4-16 is an example of an off-center exposure.

Using a grid upside down is a more common error than most technologists realize. In this case, the central portion of the beam will not be affected, but there is severe grid cutoff on the periphery of the film. The effect of using a grid upside down is much the same as that shown in Figure 4-16.

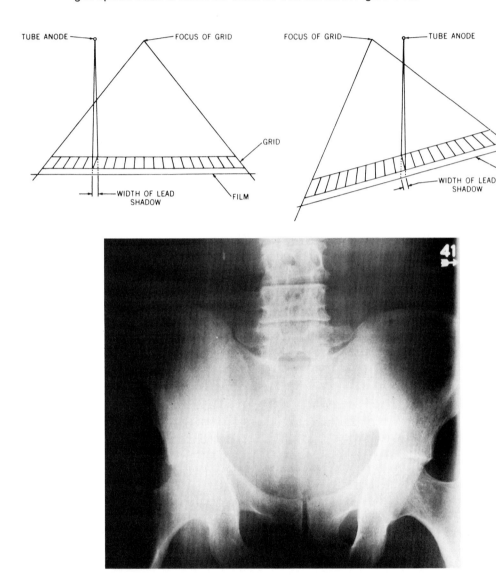

Figure 4-16. Grid cutoff due to off-center and off-level centering, but with focal distance correct. Notice how much lighter the right edge of the film is than the left. There is also a loss of detail on the left as compared with the right.

The use of grids carries with it the responsibility to correctly position the grid, to properly choose the kilovoltage for the grid being used, and to ensure that exposure of the patient to radiation is kept to a minimum. (See Table 4-2, p. 100, for suggested criteria.)

MOVING GRIDS ("BUCKY GRIDS")

Moving or reciprocating grids must continuously move during a radiographic exposure. Although the grid moves only a couple of centimeters, the effect is to remove grid lines from the radiograph. Like other grids, Bucky grids have established focal distances, and the x-ray tube must be centered over the grid to prevent off-center cutoff. Another important point to remember is that it takes a finite time for a reciprocating grid to move. If an exposure is made that is shorter than the time it takes the grid to move, the effect will be to create stop motion, and therefore grid lines will appear in the finished radiograph. With three-phase, twelve-pulse generators, exposure times may be too short for some reciprocating grids. Since the speed of movement varies so much from equipment to equipment, the technologist is advised to ascertain the shortest exposure time one may use for a particular piece of equipment from his supervisor or from the manufacturer's technical representative.

AIR-GAP TECHNIQUES

Although not a distinct mechanical means of control, the inverse square law and the angulation of the scattered photons can be used to control scattered radiation. Normally when the image receptor is placed adjacent to the object being irradiated (Figure 4-17), photons with relatively large angulations will strike the receptor; however, if the image receptor is placed some specified distance from the object, these photons will miss the receptor. Moreover, as distance increases, the energy of the photon decreases as the square of the distance. Thus, increasing the object/image-receptor distance allows a part of the x-ray photons to lose energy, and they are thus prevented from contributing to film blackening.

The air-gap technique is used only in chest radiography, cervical radiography, and occasionally in magnification angiography. There are problems with the technique. At high kilovoltages (100 kV), most of the scatter is in the forward direction, and thus photon angulation is minimal. Increasing the object/image-receptor distance does not decrease receptor-photon interaction significantly. At higher kilovoltages the energy of the scattered photon is also higher, and thus distance plays a less important role. In addition, long source/object distances are required in order to prevent image distortion due to magnification and penumbra. When applying air-gap techniques in chest radiography, one generally uses 1 inch of air gap for every foot of source/image-receptor distance (SID).

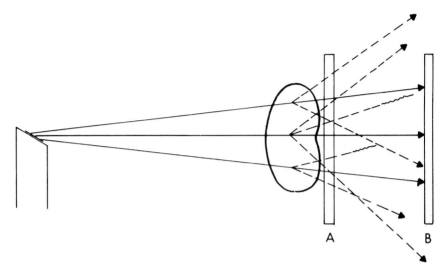

Figure 4-17. The principle of air-gap technique. If the image receptor is placed in position **A,** it collects proportionately more scattered radiation than if placed in position **B.**

SUGGESTED READING

A Home Study Syllabus on Technic for Chest Radiograph. American College of Radiology and The American Society of Radiologic Technologists. (No date.)

Janower, M. L. *Technological Needs for Reduction of Patient Dosage from Diagnostic Radiology.* Charles C Thomas, Springfield, Ill., 1963.

Lamel, D. A., Arcarese, J. S., Brown, R., and Burnett, B. M. *Correlated Lecture-Laboratory Series in Diagnostic Radiological Physics.* Bureau of Radiological Health, Public Health Service, Food and Drug Administration, Department of Health, Education and Welfare, Washington, D.C. (No date.)

Peyser, L. F. "Diagnostic X-ray Beam Collimation." *Cathode Press* 23, no. 1 (1966):38.

Walche, C., Stewart, H., and Terrill, J. "An Automatic X-ray Field Limiting System." *Radiology* 89 (1967):105–9.

REVIEW QUESTIONS

1. The primary purpose of a filter used in the diagnostic x-ray machine is

    a. To reduce contrast in the radiograph
    b. To reduce scattered radiation

c. To reduce the skin dosage

d. To increase the x-ray beam wavelength

2. The total amount of filtration (including inherent filtration) for the average diagnostic machine above 70 kV is

a. 1.5 mm of Al

b. 4.2 mm of Al

c. 3.5 mm of Al

d. 2.5 mm of Al

3. When a cylinder cone is employed, all other factors being the same, technique factors should be

a. Decreased

b. Increased

c. Unchanged

4. A grid ratio is determined by the relationship of the

a. Length to the width of the strips

b. The width to the space

c. The height to the width of the strip

d. None of the above

5. The chief purpose of a filter of aluminum placed beneath the aperture of a radiographic tube is to

a. Control the latitude of the radiographic image

b. Eliminate the light given off by the filament

c. Absorb some of the longer wavelengths

d. Filter out undesirable stem radiation

6. A radiograph is made, using a 16 : 1-ratio Bucky. The resulting film shows longitudinal streaks of uneven densities. These are probably caused by the

a. Central ray not being directed over the center of the table

b. Bucky moving unevenly during the exposure

c. Bucky stopping at some time during the exposure

d. Film moving during the exposure

7. When using a Bucky with a focused grid, if the lateral edges of the film lose density, it is an indication that the

a. Grid travels too fast

b. Focal-film distance is not correct

c. Tube is not perpendicular to the grid

d. Part-film distance is too great

8. A decrease in grid ratio would result in

a. No change in exposure

b. Increase in exposure

c. Decrease in exposure

d. Out-of-focus exposure

9. High-ratio grids, in the control of radiographic quality

a. Prevent distortion

b. Filter out 100 percent of the remnant radiation

c. Move back and forth under the x-ray table

d. Filter out a high degree of scattered radiation

e. Reduce expense or cost

10. The type of radiation called primary is that which

    a. Emerges from the tube and reaches the patient
    b. Emerges from the patient and reaches the film
    c. Is controlled by mAs
    d. Is controlled and cut off by cones

11. A radiograph is made using an 8 : 1 grid. If a 16 : 1 grid replaces the original 8 : 1 grid, to maintain the same density we need to

    a. Decrease the amount of radiation
    b. Make no change
    c. Increase the amount of radiation
    d. Use a large focal spot

12. The higher the grid ratio, the greater the

    a. Radius of the grid
    b. Speed of exposure
    c. Cleanup of the grid
    d. Focal-film distance required

13. In a grid which has lead strips 0.25 mm apart and 4 mm high, the grid ratio is

    a. 4 : 1
    b. 6 : 1
    c. 8 : 1
    d. 12 : 1
    e. 16 : 1

14. A wedge filter is used to

    a. Compensate for varying tissue densities
    b. Increase radiation to thin parts
    c. Block collimator light fields
    d. Correct alignment of grids
    e. Compensate for cone cutting

15. The Potter-Bucky diaphragm and the stationary grid are used to

    a. Localize foreign bodies in the eye
    b. Eliminate a large portion of secondary radiation
    c. Limit the diameter of the primary beam
    d. Produce radiographs free from distortion
    e. Record movements of various organs of the body

16. Cones, stationary grids, and Bucky diaphragms are similar in that their primary function is to

    a. Protect the technologist from scattered radiation
    b. Reduce distortion in the radiograph
    c. Reduce scattered radiation
    d. Reduce exposure time
    e. Permit using a longer target-film distance

17. When a 16 : 1 grid is employed, the distance that should be used is

    a. 36 in.
    b. 48 in.
    c. 40 in.
    d. 72 in.
    e. Dependent on the focal distance of the grid

18. Remnant radiation is

    a. Absorbed radiation
    b. Secondary radiation
    c. Scattered radiation
    d. Unabsorbed radiation
    e. Gamma radiation

19. Standard air-gap techniques employ

    a. 10-in. air-gap for every 1 foot of SID
    b. 2-in. air-gap for every 5 feet of SID
    c. 1-in. air-gap for every 1 foot of SID
    d. 6-in. air-gap for every 3 feet of SID
    e. 3-in. air-gap for every 6 feet of SID

20. Using a grid upside down will cause

    a. Severe grid cutoff in the center of the film
    b. Uniform lack of density
    c. Uniform increase in density
    d. Severe grid cutoff in the periphery of the film
    e. No evident change

5 ESSENTIALS OF THE RADIOGRAPH AND IMAGE FORMATION

It is important to understand the basic terms which are used in describing a radiograph. These terms include *density, contrast, detail,* and *distortion.* They will be used to describe radiographs as long as one is in the profession. One should understand and be able to define the meaning of each and to differentiate each from the others on a finished radiograph; understand the mechanisms by which these factors and subfactors are controlled; be able to manipulate the four main factors and their subfactors to produce a radiograph with optimum information.

A diagnostic radiograph should contain sufficient density and contrast to demonstrate required anatomical parts with maximum detail which is free of distortion for a given radiographic projection. There is a difference of opinion as to what degree of density is most desirable; the degree of desirable density is a matter of individual preference. There are conditions in which distortion plays an important role (for example, tomography); in other cases distortion, within reason, is of secondary importance.

DENSITY

Density, the accumulation of black metallic silver, is the general blackening of the radiograph that appears after exposure of the film and subsequent processing. To look at density in another way, we can say that it represents how radiation was attenuated by the anatomical part. Assuming optimal development of the film, density is a measurement of the quantity of radiation absorbed by the intensifying screen, converted to light, and transmitted to the film. The quantity of radiation is made up of primary, scattered, and secondary components. Variations in density may exist in a series of radiographs, and all of them may be of good quality. As a general rule, density variations should be kept to a minimum and should not vary more than 25 percent in a series of exposures of the same anatomical part. Density changes less than 25 percent are generally acceptable. Correct density is a matter of personal choice of the diagnostician; however, all densities on the radiograph should transmit some visible light.

Density, as seen in a radiograph, is affected by a number of things. Among these are the (1) thickness, (2) density and pathological changes in the part radiographed, (3) the degree of respiratory effort, (4) the distance between the x-ray tube and the part being radiographed, (5) the presence or absence of secondary or scattered radiation fog, (6) outdated film, (7) type of film, (8) chemical activity of the processing chemicals, (9) angulation and placement of the x-ray tube, (10) kilovoltage applied, (11) filtration of the x-ray beam, (12) compression binders, (13) type of radiation source, (14) method of processing, (15) type of grid, and (16) efficiency of equipment. As far as the patient is concerned, we are concerned primarily with the thickness, the atomic composition, and the specific gravity of the part to be examined.

In general terms, density refers to how light or dark a film appears. The primary factors used to control density include distance, milliamperage-time, and kilovoltage. The distance factor, which introduces the inverse square law, is rarely used to control density because of the difficulty of calculating exposure factors. For this reason almost all exposure techniques and technique charts are based on standard source/image-receptor distances, SID (formerly called focal-film distances, FFD) which are generally 91 cm (40 in.) or 183 cm (72 in.). The inverse square law will be discussed in detail in Chapter 7.

The primary factor used in controlling film density is milliamperage-second (abbreviated mAs). Milliamperage-second is calculated by simply multiplying the milliamperes by the exposure time. For example, if 300 mA is applied to the x-ray tube for a duration of 100 ms, the amount of radiation emitted from the x-ray tube would be 300 times 0.1 sec or 30 mAs. Hence, the mAs controls the quantity of radiation, that is, the amount of radiation which is to be used in forming the radiographic image. The mAs, then, is the major factor which controls the density of the radiograph, provided there is sufficient kilovoltage. Hence, mAs controls density, but not contrast.

The kilovoltage (kV) can be used to control density. However, kV also changes contrast (and density to a lesser degree), the amount of scattered radiation, and beam quality. For this reason, density changes should be effected by changing mAs rather than kV. There are a number of charts and rules of thumb for correlating changes in kV and mAs, and these will be described in Chapter 7. Other less important factors affect density and may be used to control density on the radiograph. Increasing the film development time increases density; decreasing the development time decreases density. The use of processing temperature to control density is not recommended for a number of reasons, the major one being that processing must be stabilized in an x-ray department. Changing the processing temperature also affects the contrast and speed of the film. Control of density by controlling processing temperatures is an impractical and unwise procedure.

Damage to the focal track of the x-ray tube can alter film density (Figure 5-1). This density change is caused by roughening of the focal track, thus providing a greater surface area for the electrons coming from the cathode of the tube to interact and create x-rays. The contribution of scattered radiation towards the total film density was discussed in Chapter 4.

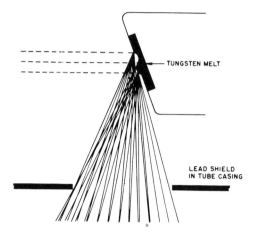

TUNGSTEN MELT

LEAD SHIELD
IN TUBE CASING

Figure 5-1. Damage to the target of a stationar x-ray tube caused by overloading the x-ray tub provides greater surface for the electrons emana ing from the cathode to strike, thus allowing mor x-rays to be produced per mAs as compared wit an undamaged target.

CONTRAST

In order to have contrast, there must be a difference between two adjacent densities as far as the transmission of light is concerned. Contrast can then be defined as "density difference." Density results from the development of black metallic silver during the processing procedure. There are varying degrees of deposition of silver; some areas on the finished radiograph may have little silver deposited; a very clear or light shade of gray would be seen when viewing these areas. On the other end of the scale, all the silver may have been deposited in an area of the film, so that after processing, this area may be entirely black. It follows that there are numerous possibilities of having different shades of gray or tone values, ranging from white to black. A tone value or shade of gray is related to the amount of silver deposition. Depending on the type of film, process-ing conditions, exposure parameters, and tissue characteristics, there are two basic types of gray scales or levels of contrast: short-scale and long-scale. *These contrast scales are controlled primarily by kilovoltage, all other things being equal.* Opinions as to the amount of contrast desirable do not vary as much as those concerning radiographic density; yet, a considerable range of opinion does exist. It is not advisable to set any degree of contrast as a standard, but as a general rule, iodinated contrast studies (such as an intravenous urogram) and bone exami-nations require high-contrast (short-scale) techniques, whereas soft-tissue exami-nations such as those of the chest, use low-contrast (long-scale) techniques.

TYPES OF CONTRAST

Short-scale (high-contrast) types are identified by a relatively few shades of gray interposed between white and black; long-scale (low-contrast) types are noted for the relatively large number of tones (shades of gray) between white and black. As contrast increases, the difference in tone value between adjacent densities becomes greater or more abrupt. The densities that represent the thinner portions of the part increase and the film may become opaque there. This increase in difference between the portion of the film that is almost totally opaque and that which is almost clear will be excessive if the densities which represent thick parts become so low that not enough density difference between structures is available to visualize detail. For example, when low kilovoltage is used for thin parts, high contrast results between the radiographic images of the bones and flesh. However, as contrast increases, a point may be reached where radiographic detail in the thinner portions, such as the subcutaneous tissues and muscles, is obliterated by opaque silver deposits; and bone detail may be lost because of the denser bone absorbing so much radiation (because the photon energy is not high enough to penetrate the bone) that little remnant radiation reaches the film. Visibility of detail in those areas is diminished, and short scale or high contrast results (Figure 5-2). As radiographic contrast decreases, the number of gray scale or tones increases, and the brightness differences in tone value between densities becomes less; consequently, more tones are visualized over the entire image. The transition between tones is more gradual, since only small differences in density occur. Figure 5-3 is a step-wedge comparison between the two basic types of contrast.

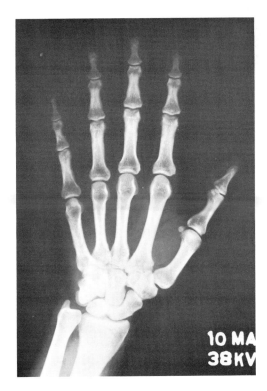

Figure 5-2. This is an example of short-scale contrast. The radiograph shows mostly blacks and whites, with few intervening shades of gray. The soft tissues are almost invisible, owing to the heavy deposit of black metallic silver.

10 MA
38 KV

Figure 5-3. Step-wedge comparison betwee short-scale (left) and long-scale contrast (righ The short scale shows only a few steps betwee black and white, whereas the long scale show many more steps.

Short-scale contrast

Short-scale contrast results when low kilovoltages are used in exposing the radiograph. The range of densities in the radiograph is short; there is a large change in tonal value from one density to another. Films exhibiting short contrast have instant eye appeal. Except for special circumstances, such as mammography, determination of fractures, and iodinated contrast studies, a radiograph with short-scale contrast is incomplete because details that represent the thinnest and thickest portions of the body structure are not always shown. This leads one to the conclusion that short-scale radiographs generally offer less information than films of longer scale contrast. Typical examples of short-scale contrast are shown in Figure 5-4 (left view).

Long-scale contrast

Long-scale contrast makes possible the visualization in a single radiograph of a wide range of tissues with different absorption efficiencies. The shorter wavelengths afforded by higher kilovoltages allow greater penetration of tissues and

114

result in an abundance of remnant radiation of varying intensities, which in turn produces a larger number of tones. Long-scale contrast, within reason, is the most useful contrast scale in medical radiography. Typical examples of long-scale contrast are shown in Figure 5-4 (right view). The student should be cautioned that the contrast scale should never be so long that differentiation between structures is difficult (see Figure 5-5).

For some time there has been disagreement over radiographs with short-scale versus those with long-scale contrast. Some believe that all radiographs made with high kilovoltages are necessarily gray and flat. But it is as possible to produce films of good quality with high kilovoltages as with low kilovoltages. Radiographs made at various kilovoltage values, but with suitable compensation in mAs, should be adequate in quality at different contrast levels. Optimum contrast is that which provides the most information for a given examination. When a choice is to be made, the choice should always be in favor of the higher kilovoltage techniques, since these result in less exposure of the patient to radiation.

Figure 5-6 shows two exposures of the toes, made on nonscreen film. The anteroposterior projection was exposed at 50 kV and 25 mAs; for illustrative purposes, the lateral exposure was made at 100 kV, 10 mAs. Note the long range of densities seen at the higher-kV exposure. Figure 5-7 illustrates two exposures of the os calcis made on the same nonscreen film, one exposure at 50 kV, 60 mAs, and one exposure at 120 kV, 15 mAs. Note that these exposures are comparable.

Radiographic contrast

Radiographic contrast results from the interaction of two main factors: (1) film contrast, which is inherent in film design and manufacture and is optimized by the developing process, and (2) subject contrast, which is affected by the differential absorption of the radiation by different areas of the body part being examined. Radiographic contrast can be controlled by altering either one or both of these contributing factors. Film contrast can be changed by changing films or processing techniques; however, the film and development should be standardized, and the control of contrast in the finished radiograph should be left to a single factor—kilovoltage. Subject contrast can be altered readily by changing the kilovoltage, which in turn affects the quality of radiation.

Screen contrast

Textbooks such as this have not often included the contributions that intensifying screens make to the levels of contrast seen in a radiograph. In earlier days when most intensifying screens were manufactured from calcium tungstate, the absorption of the x-ray beam by the screen was insignificant for changing beam quality (and thus contrast). However, with the advent of rare earth and other newer phosphors for intensifying screens, the degree of photon absorption by the screen is 2 to 3 times as great as medium-speed calcium-tungstate screens.

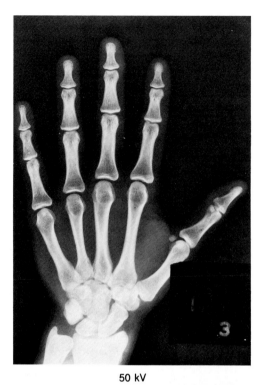 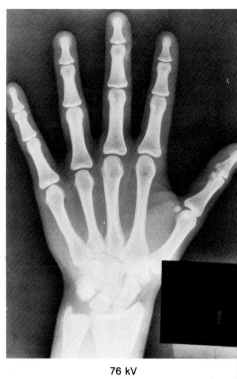

50 kV 76 kV

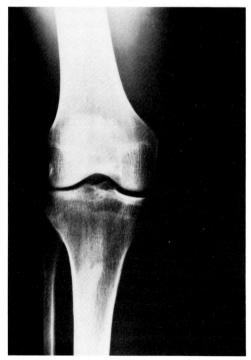 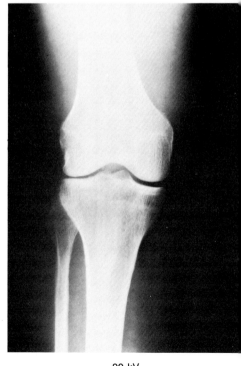

60 kV 90 kV

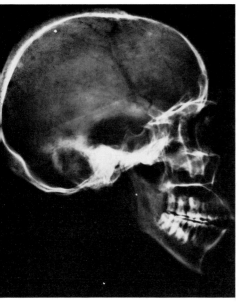
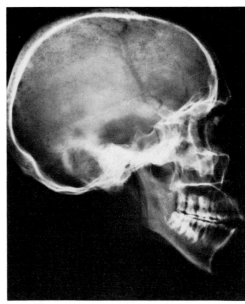

70 kV   92 kV

Figure 5-4. Comparison between short-scale and long-scale contrast in matched-density exposures. The short-scale exposure is shown on the left, its corresponding long-scale exposure on the right.

This increased absorption may change the quality of the x-ray beam, and there is a possible measurable difference in contrast caused by the screens. The contribution to the contrast by intensifying screens varies so much that no hard-and-fast rule can be given. However, as a general rule, some rare earth and other newer phosphors give a higher-contrast film than comparable calcium tungstate screens when using non-grid techniques.

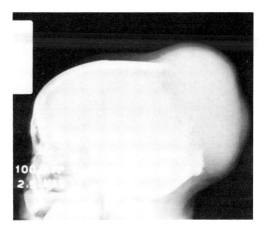

Figure 5-5. Extreme long-scale contrast. This exposure was made in order to visualize skin lines, with 110 kV and 2.5 mAs. The exposure has the appearance of a very short-scale, underexposed radiograph because of the obliteration of detail caused by gross overpenetration of the object. Films of this nature are useful only for special purposes.

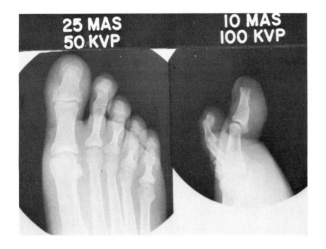

Figure 5-6. Two views of phalanges with exposures at different kilovoltages and appropriate compensation in mAs. Non-screen exposure.

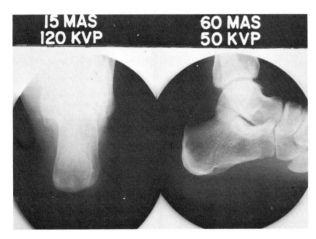

Figure 5-7. Two views of the os calcis made with high and low kilovoltages and appropriate change in mAs. Non-screen film.

CONTRAST AND BEAM QUALITY

Of all the primary factors necessary for the exposure of a radiograph (milliamperage, time, kilovoltage, and distance), kilovoltage has the greatest effect upon the radiographic image; it influences the quality of the radiation, determines the contrast in the image, influences exposure latitude, provides secondary control of radiographic density, and affects secondary or scattered radiation fog. When secondary and scattered radiation and mAs are controlled, kilovoltage becomes the controlling factor of subject contrast. Contrast enhances detail in the image; thus, detail in the image is dependent upon subject contrast, which in turn is controlled by the kilovoltage selected.

Radiographic penetration

In order to achieve acceptable contrast, the part being examined must be adequately penetrated by kilovoltage. By definition, penetration is the act of something passing into or through something else. There are various degrees to which one thing can penetrate something else; penetration can be minute, partial, or complete. The same thing holds true in radiography; an x-ray beam can be absorbed totally (no penetration to the image receptor), partially, or not at all (complete penetration to the image receptor). The degree to which an x-ray photon penetrates matter is related to its wavelength, which in turn is related to its energy level. The shorter the wavelength, the higher the photon energy. In radiography, we would like most of the x-ray photons to possess enough energy to penetrate the object being irradiated, and thus contribute to the formation of the radiographic image. Figure 5-8 illustrates this concept.

The energy level, also called the quality, of the x-ray photon is determined by the kilovoltage applied to the x-ray tube. (We ignore here the contributions of filtration toward beam quality. The higher the kilovoltage applied to the tube, the higher the penetrating capability of the x-ray beam. In formulating x-ray techniques and as a rule of thumb, one must establish that the object being irradiated has been adequately penetrated before determining the appropriate level of film blackness as required by the contribution of the quantity of x-ray photons. If the penetrating power of the x-ray beam is insufficient, most of the x-ray image will be formed by scattered or secondary radiation, with its inherent loss of detail (see Chapter 4).

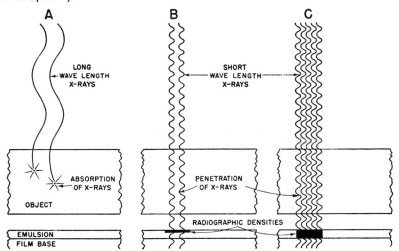

Figure 5-8. The penetrating power of x-rays. Scheme **A** shows x-rays of long wavelength being absorbed by the object. in **B,** x-rays of short wavelength penetrate the object and expose the film (remnant radiation). In **C,** increasing the number of x-rays by the mAs factor increases the number reaching the film, and more silver in the film is exposed than in **B.**

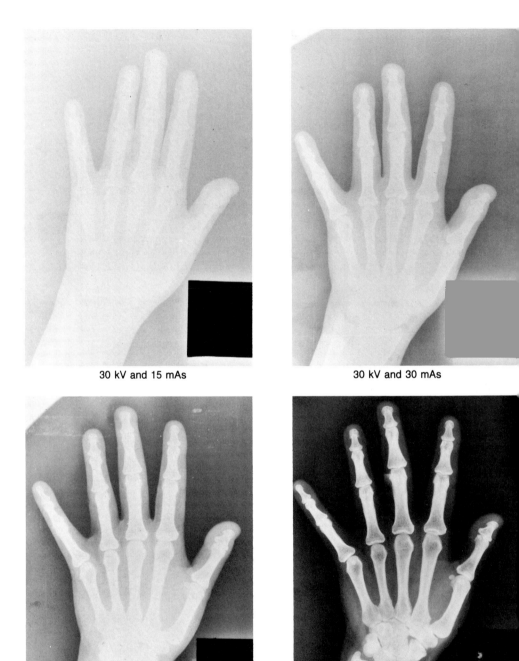

30 kV and 15 mAs

30 kV and 30 mAs

30 kV and 60 mAs

30 kV and 240 mAs

Figure 5-9. Exposures of a hand phantom with kilovoltage kept constant and the milliamperes-seconds increased. Although the density of the film increases, the trabecular pattern of the carpal bones still cannot be seen clearly.

Figure 5-9 illustrates this very basic fact. In this case a radiograph was made of a phantom at a low kilovoltage. Subsequent exposures were made by increasing the mAs for each exposure. As can be noted in the illustration, although the density or blackness of the film increases with increasing mAs, one is still unable to see clearly the bone trabeculation in the carpal bones. Figure 5-10 is a repeat of this experiment, but in this case an adequate kilovoltage has been chosen. In the first exposure, although the density is less than desirable, the trabecular pattern with the carpal bones can be seen. The second exposure shows both proper penetration and proper density. This experiment shows that *no practical quantity of radiation (mAs) can compensate for an inadequate level of x-ray photon energy (kV)*.

As the student proceeds through his study of radiologic technology, he should develop and fix in his mind the basic kilovoltages that must be used for optimal penetration of particular body parts. Since this concept of adequate penetration is extremely important in formulating x-ray techniques, the student should repeat the above experiments using different types of phantoms, screens, and films. Adequate penetration is present when one can "see through" the densest area of the part being radiographed.

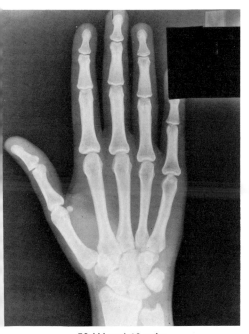
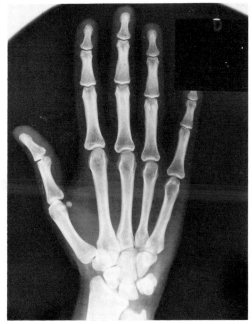

50 kV and 10 mAs

50 kV and 50 mAs

Figure 5-10. A repeat of the experiment shown in Figure 5-9. The film on the left is underexposed, but adequately penetrated. The film on the right is correctly exposed and penetrated. Note how well one can see the trabecular pattern in the carpal bones as compared with Figure 5-9.

Control of contrast with kilovoltage

There is an old rule in radiography: "the lower the kilovoltage, the higher the contrast." The reverse also holds true: "the higher the kilovoltage, the lower the contrast." These rules are valid only within certain limits. If the kilovoltage is too low, i.e., below 35 kV, there may not be a sufficient film density to allow for adequate contrast, or the contrast may be too great, with the result that detail is obscured. With diminished density, the degrees of brightness (tone values) over the radiograph are too nearly the same for satisfactory discrimination between the density levels. To have sufficient contrast, one must also have adequate density. As a general rule, contrast changes are affected by kilovoltages in a range between 35 and 90, with greater contrast changes per unit change in kilovoltage seen as the 35-kV level is approached, and less as the 90-kV level is approached. Above and below these levels, there is no significant change in contrast as the kilovoltage is changed.

Exposure errors with kilovoltage

Technically we should like to dispose of the terms "overexposure" and "underexposure" applied to kilovoltage. We should assume that the radiograph has been correctly exposed with the appropriate kilovoltage and mAs to obtain a proper film density. When a radiograph has been exposed at a kilovoltage that is too high for the part being examined, the finished product is dull and gray; the contrast scale is too long, and image detail is diminished. This dull look of the radiograph is due to the excessively long gray scale, the increased forward scattering of x-ray photons, and subsequent film fogging. Under normal circumstances x-ray photons produce scatter and secondary x-rays which are emitted in all directions. As kilovoltage is increased, there is a greater proportion of scatter x-rays emitted in the forward direction than sideways. Thus as kilovoltage is increased, more scatter strikes the image receptor. Usually, with an optimum kV, a reduction of 10 kV with a 2× increase in mAs will improve radiographic quality appreciably. When using high kilovoltage, film quality can also be improved by inserting a grid in front of the image receptor. Inadequate kilovoltage brings about underpenetration, which occurs when there is insufficient kilovoltage to penetrate the object being irradiated. There must be a minimum level of kilovoltage to allow an adequate penetration. It is unsafe and unwise to use excessive mAs to compensate for inadequate kilovoltage penetration, for the reasons already mentioned. Exposures made at unreasonably low kilovoltages (assuming adequate penetration) produce films that are mostly blacks and whites with few shades of gray in between.

Kilovoltage and exposure latitude

Exposure latitude, sometimes called technique latitude, is the amount of error that one can make in exposing a radiograph and still produce a diagnostically acceptable film. Thus, it is the difference between the maximum exposure that can be used to produce a diagnostically acceptable radiograph and the minimum exposure that would produce an acceptable radiograph. There are two factors to consider in exposure latitude: (1) the amount of density and contrast that is placed on the film by the combination of kV and mAs, and (2) the response of the image-receptor system to exposure.

It has been widely held that if high-kilovoltage techniques are used, there is a wider margin for error in choosing the appropriate mAs for an acceptable film density than at lower kilovoltages; that is, there is a wider margin for error at higher-kilovoltage techniques than at lower-kilovoltage techniques. The statement is true, but for the wrong reason. The effect is due, not to a wider margin for error in choosing mAs, but to the higher penetrating power obtained at the higher kilovoltage levels. At these higher kV levels, the high kV/low mAs combination will allow a proper exposure for a greater variation in part thicknesses. For example, if a low kV-mAs exposure will produce an adequate film for a part thickness of 6–8 cm, a matched-density exposure at high kV levels will allow an adequate exposure for part thicknesses measuring perhaps a range of 5 to 9 cm.

That the view that there is greater technique latitude at high kilovoltages due to mAs is a misconception can easily be proved (see Figure 5-11). An exposure of a phantom is made at 67 kV, 60 mAs, with medium-speed calcium tungstate screens, and a matched-density exposure is made at 96 kV, 20 mAs. Additional exposures were made at each kV level by first decreasing and then increasing mAs by 50 percent, i.e., 30 mAs, 90 mAs, and 10 mAs, 30 mAs, respectively. Compare the exposures for density and note how the film density at each kV is matched with its corresponding film. Thus on changing the 96-kV, 20-mAs exposure to 96-kV, 10-mAs, the film density is the same as the exposure made at 67 kV, 30 mAs. The percentage change of mAs at each change of kV level will produce the same change in film density.

Exposure latitude, as far as kilovoltage is concerned, becomes wider as higher kilovoltages are used, and there is less chance of having to repeat an exposure because of an error in measuring part thickness or calculating exposure factors. This exposure latitude is due to the greater penetrating power of the kilovoltage rather than a greater latitude in choosing mAs. The trade-off is that higher kilovoltages provide a longer scale of contrast—which may or may not be desirable, depending on the likes or dislikes of the viewer. This increase in exposure latitude is one of the reasons why many x-ray departments use high-kV techniques in chest radiography (the major reason is the need for a long gray scale film to detect subtle changes within the lung). Not only does the use of higher-kV techniques allow more margin for error; it also subjects the patient to less radiation and decreases the number of repeat examinations.

The speed of the recording system has become a major concern with the advent of the high-speed recording systems. There are new recording systems that are up to many times as fast as the calcium tungstate reference system. If an exposure of 1 milliroentgen (mR) is required to produce a net density of 1 for the reference system (medium-speed calcium tungstate screen/film), then a

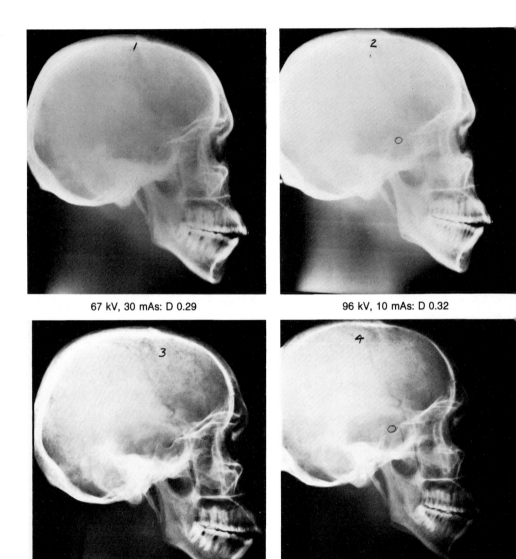

67 kV, 30 mAs: D 0.29

96 kV, 10 mAs: D 0.32

67 kV, 60 mAs: d 0.62

96 kV, 20 mAs: D 0.69

system eight times as fast would require one-eighth of that exposure, 0.125 mR, to produce the same film density. Although the percentage change for the radiation required to produce identical density is the same for both the standard and high-speed systems, the actual number changes (for example 100 mAs to 12.5 mAs) indicate a very narrow margin for error. In radiography, a density that is plus or minus 25 percent of an ideal density for a particular examination will generally produce a diagnostically acceptable radiograph. If an examination using the medium-speed calcium-tungstate system requires 100 mAs, the use

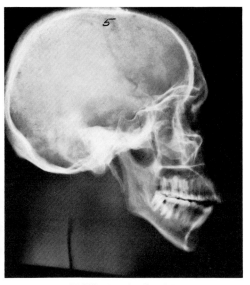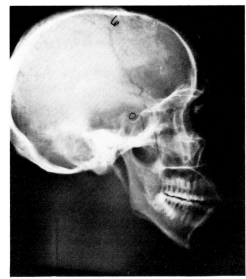

67 kV, 90 mAs: D 1.07   96 kV, 30 mAs: D 1.09

Figure 5-11. Matched-density exposure at high and low kilovoltage. Additional exposures were made by changing mAs by ± 50 percent. Exposure factors and density readings are noted for each exposure. Notice that the percentage change in density is the same for each level of kilovoltage and subsequent change in mAs.

of 75 or 125 mAs will normally produce an acceptable radiograph. On the other hand, if a high-speed system which is eight times as fast as the calcium-tungstate system is used, our required range is now between 9 and 15 mAs. The new, very-high-speed systems therefore require much more meticulous attention to be paid to measurement of part thicknesses and calculating exposure factors.

The unfortunate thing about the very fast recording systems is that the mAs has to be maintained at its pre-high-speed level in order to control quantum mottle. Therefore, when converting from a medium-speed to a high-speed recording system, kilovoltage rather than mAs is lowered. Lowering the kilovoltage lowers the exposure latitude (margin for error), and at the same time selection of mAs becomes up to eight times as critical. This is to say that when the technologist uses high-speed recording systems, he must take care to measure accurately the part to be radiographed and carefully check the technique charts posted in the exposure control area.

DETAIL/DISTORTION

Detail in a radiograph refers to how well one can see small structures. When structures have sharp borders and are clearly defined, and when the density differences between structures are sufficient for the eye to distinguish one structure easily from another, the radiograph is said to have good detail. Sharpness of detail is often referred to as "definition." In order to have detail, the part being radiographed must be adequately penetrated by kilovoltage and there must be an acceptable level of density, as has already been discussed in this chapter. Since detail is related to distortion and vice versa, the two will be discussed together.

Detail in a radiograph is dependent primarily on geometric factors and factors that affect visibility. These factors are categorized in Table 5.1.

Scattered radiation and its control, screen speed and resolution, and screen/film contact have been discussed elsewhere in this text and will not be discussed here. Factors affecting detail—overexposure and underexposure of the x-ray film, use of filters, fogging of film for any reason, improper use of grids, pathological involvement of the part to be radiographed, and the characteristics of the tissue being examined—are discussed elsewhere in the text and will not be covered here.

Focal spot size

The effect of focal spot size on radiographic detail has often been ignored in the past. The reason why is that in the past there were few x-ray tubes from which to choose and x-ray generators were of limited power output. With the introduction of large x-ray generators, a greater selection of x-ray tubes having higher heat ratings, along with new technology and new procedures, became available, and the importance of the focal spot on detail recording could no longer be ignored. (For more detail information on focal spots of x-ray tubes, the reader is referred to the work of L. M. Bates in Suggested Reading, at the end of this chapter.)

X-ray tubes are generally supplied with a large and a small focal spot. The large focal spots are used for high-load (high-mA) exposures—generally those with milliamperage over 200 or 300. Small focal spots are used for detail work. Large focal spots are generally supplied in sizes of 1.0, 1.2, 1.6, or 2.0 mm; small focal spots are generally 0.3, 0.6, or 1.0 mm. The best combination of

Table 5-1. Factors affecting detail

Geometric	Visibility
1. Focal spot size	1. All radiographic exposure
2. SID	and processing factors
3. OID	
4. Motion unsharpness	
5. Type of film	
6. Type and speed of intensifying screens	
7. Screen-film contact	

large and small focal spots, taking into consideration the instantaneous loading capability of the x-ray tube and field coverage, is a 0.6/1.2; i.e., a 0.6-mm small focal spot and a 1.2-mm large focal spot. The generator console indicates to the operator which focal spot is being used.

The size of the focal spot of the x-ray tube has a great effect on recording detail. Generally, the smaller the size of the focal spot, the sharper the detail. It is commonly stated that a focal spot cannot resolve anything smaller than itself, and this is generally true. Focal spot size is not constant; it broadens (blooms) with decreasing kilovoltage and increasing mAs. Changing the mA station from a 300 to a 1000 (large focal spot) will cause the focal spot to broaden owing to the increased current supplied to the cathode filament; the hotter the filament becomes, the larger it will be. For this reason it is wise to use the smallest mA station which will allow an adequate exposure time for the part being examined and still allow recording of good detail.

With kilovoltages generally less than 50 kV, the focal spot also broadens, but for a different reason and to a lesser extent. When kilovoltage is applied to the x-ray tube, the voltage differential between anode and cathode draws electrons emitted from the cathode towards the anode, where they strike the focal track of the x-ray tube. The higher the voltage differential, the straighter the path of the electrons flowing from the cathode to the anode. At low kilovoltages, below 50 kV, these electrons tend to drift slightly apart and therefore do not strike the focal track in as compact a fashion as would be desirable. This spreading apart of the electron beam at the focal track effectively enlarges the focal spot. There is negligible effect on broadening for the focal spot due to kilovoltage with kilovoltages greater than 50.

Geometric unsharpness

Lack of sharpness implies decreased detail. If sharpness is decreased, distortion is a contributory factor. Geometric unsharpness occurs when the part to be radiographed is magnified and distorted. It results from a combination of focal spot size, source/image-receptor distance (SID) and object/image-receptor distance (OID).

For all practical purposes, the size of the focal spot has little to do with detail when the SID is greater than 183 cm (72 in.). At distances less than this, focal spot size becomes important. The contribution of focal spot size to image distortion may be understood better by considering the focal spot as a light source of finite size. When an image is projected by using a small powerful light source such as a spotlight (i.e., a very small focal spot), the borders of the projected image are sharp. On the other hand, if the light source is large, like a flood lamp (large focal spot), the image is ill-defined at the outer edges because of the contribution of penumbra. Penumbra results from the undercutting of the image by light photons emanating from different points within the light source. It follows that in order to obtain maximum radiographic detail from the focal spot, penumbra must be reduced (Figure 5-12). Penumbra is reduced by using

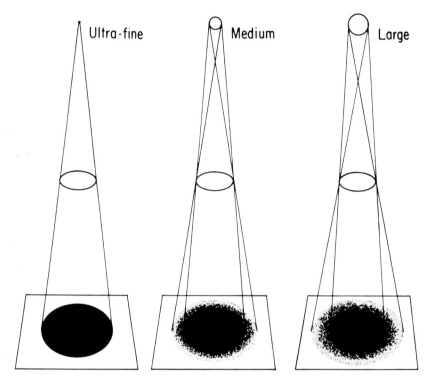

Figure 5-12. Effect of focal spot size on penumbra. As the illustration shows, the smaller focal spot size, the sharper the detail.

the smallest focal spot possible and the longest SID possible, but keeping in mind the heat-loading characteristics of the x-ray tube.

Just as important as are the focal spot size and the SID in controlling penumbra, is the OID—that is, how close the object being irradiated is to the recording surface. Suppose, as in Figure 5-13, that there is light from a point L falling on a white card C, and an opaque object O is interposed between the light source and the card. A shadow of the object will be formed on the surface of the card. This shadow will naturally show some enlargement because the object is not in contact with the card; the degree of enlargement will vary according to the relative distances of the object from the card and from the light source. The form of the shadow also may differ according to the angle which the object makes with the incident light rays. Deviation from the true shape of the object as exhibited in its shadow image is called *distortion*.

Figure 5-13. The principle of geometric unsharpness: the effects of changing the relative positions of the source, object, and card (image receptor). **A.** A very small focal spot at a long SID allows sharp definition. **B.** A large focal spot (illustrated as two point sources) and short SID causes enlargement and blurring of the object. **C.** Increasing SID greatly decreases distortion. **D.** Penumbra is decreased by placing the object closer to the image receptor. Tilting either the x-ray tube **(E)** or the object **(F)** causes distortion of the image. (Courtesy, Eastman Kodak Company.)

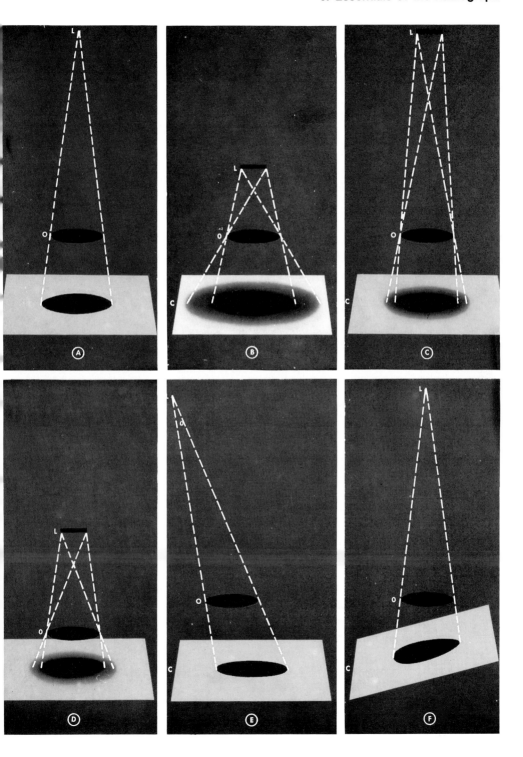

The degree of sharpness of any shadow depends on the size of the source of light and on the position of the object between the light and the card—whether nearer to or farther from one or the other. The shadows cast, when the source of light is not a small area, are not perfectly sharp, because each point in the source of light casts its own shadow of the object; all these overlapping shadows, since they are slightly displaced from one another, produce an ill-defined image. Figure 5-14 shows the effects of changing the relative positions of source, object, and card.

The penumbra produced by the focal spot as related to SID and to OID causes enlargement of the radiographic image. The geometric enlargement decreases sharpness; it is commonly called *geometric unsharpness.* To control geometric enlargement, one must use the longest SID possible and place the object as close to the image receptor as possible.

Motion unsharpness

As the term implies, motion unsharpness occurs when the part being radiographed is moved during the time of the exposure. It occurs when the patient breathes during the time of the exposure, or when he moves the part being radiographed, either voluntarily or involuntarily. This type of motion unsharpness can be controlled if the technologist will watch the patient and use as short an exposure time as possible. Motion unsharpness produces a blurring of the radiographic image somewhat like that which happens when one moves a camera while an exposure is being made (Figure 5-15).

The contribution to image degradation of motion unsharpness caused by vascu-

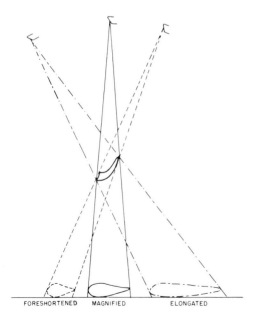

FORESHORTENED   MAGNIFIED   ELONGATED

Figure 5-14. Types of distortion caused by inaccuracies in aligning the tube, the object, and the film. If the tube is angled to the side of the object, the projected image will be either foreshortened or elongated. This type of distortion is called disproportional magnification. If the alignment is correct, but the object is placed away from the image receptor, the image will be magnified, which is a form of proportional distortion.

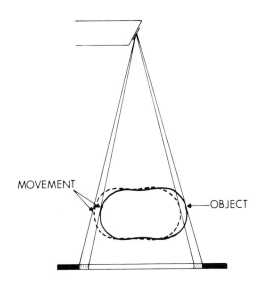

MOVEMENT

OBJECT

Figure 5-15. The effect of motion on detail. As the body part moves during the time of exposure, blurring of the object occurs and small detail is obscured.

lar pulsations and peristalsis can be somewhat controlled by knowing the minimal exposure times that one can anticipate for a specific type of examination. Although it is not necessary to go into detail concerning the derivations of these exposure times, it is important for the radiologic technologist to know what the minimal exposure times are for radiography. These times are as noted in Table 5-2.

SUMMARY

To summarize, it is important to understand the basic descriptive terms which are used in describing radiographs. These factors are controlled as noted below and as summarized in Table 5-3.

1. Kilovoltage controls penetration by controlling the wavelength of photons. No practical quantity of mAs can compensate for inadequate penetration.
2. When all factors are constant, density changes with kilovoltage. With changes in kilovoltage, film density must be compensated for by changing mAs or time or both.

    When all exposure factors, except mAs, are constant, radiographic density increases in direct proportion to the mAs. Changes in mAs must be compensated for by corresponding changes in kilovoltage or time, all other things being unchanged.

    Density is controlled primarily by a combination of milliamperage and time.
3. When all other factors but time of exposure remain constant, radiographic

Table 5-2. Exposure time requirements for various degrees of motion[a]

Organ	Motion	Exposure Time to Limit Motion to:			
		1.0 mm	0.5 mm	0.1 mm	Optimal
Head and neck	1–5 mm/sec	1000 ms	500 ms	100.0 ms	50–100 ms
Heart	100 mm/sec	10 ms	5 ms	1.0 ms	3–7 ms
Lungs	100 mm/sec	10 ms	5 ms	1.0 ms	20 ms
Heart valves	500 mm/sec	2 ms	1 ms	0.2 ms	1–2 ms
Abdomen	50 mm/sec	20 ms	10 ms	2.0 ms	50–80 ms
Extremities	1 mm/sec	1000 ms	500 ms	100.0 ms	50 ms

[a]Exposure times required to control motion unsharpness. The major organs are listed along with the time requirements necessary to provide certain degrees of limitation of motion. Faster exposure times are required if less motion is required. An optimal exposure time is listed in the last column. (From Thompson, *A Practical Approach to Modern X-Ray Equipment*, Little, Brown and Company, Boston, 1978.)

Table 5-3. Summary of radiographic and image formation

Photographic aspects	Geometric aspects

I. Radiographic density:
Controlled by
  milliamperes-seconds
  source-image receptor distance
  anode heel effect
Influenced by
  kilovoltage
  fog
  processing
	I. Image sharpness (definition):
	Affected by
	  focal spot size
	  object/image receptor distance
	  source/image receptor distance
	  motion
	  screen speed
	  film/screen contact

II. Radiographic contrast:
Controlled by
  kilovoltage
Influenced by
  beam filtration
  fog (all forms)
  development
  tissue contrast
  type of film
  pathology
	II. Image size (magnification):
	Influenced by
	  Source/image receptor distance
	  Object/image receptor distance

III. Secondary radiation fog:
Affected by
  size of area irradiated
  compression of part
  cones and collimators
  diaphragms
  grids
  kilovoltage
	III. Image shape (distortion):
	Influenced by
	  alignment of tube to film
	  alignment of part to film
	  object/image receptor distance

density increases in direct proportion to time within the framework of 10 ms–2 sec. (See Chapter 2 concerning reciprocity failure.)

4. The primary controlling factors for radiographic contrast are inherent film contrast and subject contrast. The primary controlling factor for subject contrast is kilovoltage.

   As kilovoltages change, subject contrast changes; as kilovoltages increase, subject contrast decreases; and vice versa. The lower the kilovoltage, the shorter the scale of contrast; the higher the kilovoltage, the lower the scale of contrast.

5. Detail in a radiograph is dependent primarily on contrast levels, focal spot size, source/object/image-receptor distances, motion unsharpness, collimation, and the characteristics of the recording system.

   The larger the focal spot, the less the detail that can be recorded, and vice versa. Focal spot size increases with increasing milliamperage and decreasing kilovoltage.

6. Geometric unsharpness is contributed to by SID and OID and is controlled by using the smallest focal spot possible for a designated examination, by using SIDs that are as long as possible, and by keeping the part to be radiographed as close as possible to the image receptor.

7. Motion unsharpness is caused by voluntary or involuntary movement of the part being radiographed during the time of exposure. Motion unsharpness is decreased by decreasing exposure times.

SUGGESTED READING

Bates, L. M. *Characteristics of Radiographic Films and Screens.* Physics of Diagnostic Radiology Proceedings. USDHEW Publication No. (FDA) 74–8006, 1973.

Bates, L. M. "Some Physical Factors Affecting Radiographic Image Quality: Their Theoretical Considerations and Measurement." *Environmental Health Ser. (Radiol. Health)* 38, no. 1 (1969).

Buchignani, J. S., and Howlett, P. "Changes in the Latent Image of X-ray Film." *Radiology* 108 (1973):213–15.

Cleare, H. M. *General Physics of X-ray Film-Screen-Process Speed and Contrast Characteristics.* Proceedings of Symposium on Medical X-ray Photo-Optical Systems Evaluation. USDHEW Publication No. (FDA) 76–8020, 1974.

Doi, K., et al. "Effect of Geometric Unsharpness upon Image Quality in Fine-Detail Skeletal Radiology." *Radiology* 113 (1974):723.

Lamel, D. A., Arcarese, J. S., Brown, R., and Burnett, B. M. *Correlated Lecture-Laboratory Series in Diagnostic Radiological Physics.* Bureau of Radiological Health, Public Health Service, Food and Drug Administration, Department of Health, Education, and Welfare. Washington, D.C. (No date.)

Milne, E. N. C. X-Ray Tube Focal Spots. In *Medical X-ray Photo-Optical Systems Evaluation.* DHEW Publication (FDA) 76–8020. U.S. Department of Health, Education and Welfare, Public Health Service, Food and Drug Administration, October 1975.

Rossman, K. "Image Quality." *Radiol. Clin. of N. Amer.* 7 (1969):419.

Yuen, K., et al. "Physical Factors Relating to Detail on a Radiograph." *J. Can. Assn. Radiol.* 15 (1964):202.

REVIEW QUESTIONS

In the questions below, choose the *best* answer.

1. The sharpness of detail in a radiograph is dependent on all of the below EXCEPT
   a. Size of the x-ray generator
   b. Size of the focal spot
   c. SID
   d. Immobilization
   e. OID

2. Radiographic contrast is the result of all of the below EXCEPT
   a. Focal spot used
   b. kV used
   c. Volume of tissue being irradiated
   d. Density of tissue being irradiated
   e. Type of film used

3. The factor which produces the *greatest* unsharpness is
   a. Focal spot size
   b. Long SID
   c. Motion
   d. High-speed screens
   e. High-speed film

4. The sharpness of detail in a radiograph is decreased by the use of
   a. A large focal spot
   b. A high mA station
   c. High-speed calcium tungstate screens
   d. A and B are correct
   e. A, B, and C are correct

5. Short-scale contrast results primarily from
   a. High filament current
   b. Nonscreen film
   c. Use of high kilovoltage
   d. Secondary radiation
   e. None of the above

6. Subject contrast is chiefly influenced by the
   a. Milliamperes-seconds (mAs)
   b. Kilovoltage
   c. Milliamperage
   d. Time
   e. Type of film used

7. The wavelength of an x-ray beam is determined by
   a. The quantity of electrons in the cathode stream
   b. The milliamperage
   c. The voltage in the filament circuit
   d. Voltage differential between anode and cathode
   e. The size of the focal spot

8. Geometric unsharpness is controlled *primarily* by

    a. Decreasing OID
    b. Increasing the size of the focal spot
    c. Increasing the SID
    d. Decreasing exposure time
    e. Increasing the speed of intensifying screens

9. Motion unsharpness is controlled primarily by

    a. Suspending respiration
    b. Decreasing exposure time
    c. Increasing the speed of the film/screen system
    d. A and B are correct
    e. A, B, and C are correct

Choose from Column B the item that matches the item in Column A. Each item in Column B may be used once, more than once, or not at all.

Column A	Column B
10. ____ Control of penetration	a. Time
11. ____ Control of contrast	b. Kilovoltage
12. ____ Control of density	c. Milliamperes-seconds
13. ____ Quantity of radiation	d. Focal spot size
14. ____ Quality of radiation	e. mA station
15. ____ Scattered radiation	

Questions 16–24. In the chart below, a standard exposure has been defined. Subsequent changes have been made in the exposure parameters to produce desired changes in the radiograph. For each question, select an answer from the following:

    a. Increased
    b. Decreased
    c. No change
    d. Change expected

		kV	mAs	Screen	OID	SID
	Standard	70	30	1	10 cm	100 cm
16. ____	Film density	70	15	2×	NC	NC
17. ____	Film density	70	50	1	NC	NC
18. ____	Film density	78	NC	NC	20	100
19. ____	Contrast	78	15	2×	15	NC
20. ____	Contrast	62	30	.5×	10	80
21. ____	Contrast	90	5	NC	20	100
22. ____	Detail	85	7	.5×	30	100
23. ____	Detail	70	30	3×	15	150
24. ____	Detail	65	40	.5	10	100

kV = kilovoltage
mAs = milliamperes-seconds
Screen = medium-speed, with speed equal to 1
OID = object/image-receptor distance
SID = source/image-receptor distance
NC = no change

6 BASIC EXPOSURE EXPERIMENTS

Up to this point, we have been concerned with learning the principles that underlie exposure techniques. This chapter is designed to allow "hands on" experience with x-ray equipment. Thus, we can apply what we have learned in a modified work setting. At the outset, the student should be familiar with all the controls in the radiographic room. Since there is so much variation in the controls from one room to another, it would be advantageous if a tour of the x-ray department could be arranged, in order to take note of the following factors:

1. Target angle, target diameter, and focal spot size of each x-ray tube. The proper tube rating chart for each x-ray tube should be reviewed, and the instructor should make sure that the student knows how the rating chart should be used. The high-speed rotor control (if available) should be identified and the student instructed in its proper use.
2. Locate and identify the types of cassettes, film holders, screens, and films that are used in the department.
3. Ascertain that radiographic processing is stabilized and optimized.
4. Locate and understand the use of the generator exposure controls;
   a. Exposure control, button versus hand-switch, single-position versus two-position switch.
   b. Major and minor kilovoltage controls.
   c. Location of Bucky control switch and how it operates.
   d. Major power disconnect for the room, and how it operates.
   e. Generator capacity; single-phase, three-phase, 2, 6, or 12 pulses, constant potential, falling load.
   f. Milliamperage selection control and how it is operated.
   g. Timing controls; decimal versus fractional system.
   h. Collimator/cone light-field alignment with radiation-field size.
   i. Source/image receptor distance indicator.
   j. Tube angulation factor.
5. Availability of phantoms and step wedges.
6. Technique chart availability and accuracy.
7. Thickness indicator.

Even if an energized teaching laboratory is used, the student should still go through the process noted above and determine for himself what an x-ray room is capable of doing, if for nothing more than to develop good working habits.

No technologist should ever expose a patient to ionizing radiation until he is familiar with the capability of the x-ray equipment being used; otherwise a patient is needlessly exposed to radiation. If there is ever any question, exposures should be first made of phantoms rather than of human subjects. The experiments in this chapter have been designed to use a radiographic phantom. Under no circumstances should one be allowed to perform these experiments on patients, fellow students, or instructors. By performing these experiments, the technologist will acquire a knowledge of radiographic technique that cannot be gained in any other way.

Unless otherwise stated it is assumed that the student will routinely collimate to the area of interest, that no filtration is added to the x-ray above the required 2.5 mm aluminum inherent/added filtration, and that optimized automatic processing is available.

Experiment 1. The effect of milliamperes on film density with all other factors constant.

Purpose. To show that film density changes in proportion to tube current (mA).

Theory. Any change in tube current will affect the quantity of x-rays emitted from the tube. As the milliamperage to the x-ray tube is increased or decreased, the cathode filament is correspondingly heated or cooled. The greater the milliamperage applied to the cathode filament, the hotter the filament becomes and the greater the quantity of x-rays emitted from the tube.

Procedure. Make 3 exposures using the following factors:

Absorber:	Hand phantom
Film:	Any 24 × 30 cm (10 × 12 in.)
Screens:	Medium-speed calcium tungstate
Grid:	None
Distance:	100 cm (40 in.)
kV:	50 (single-phase)
Time:	100 ms ($\frac{1}{10}$ sec)
mA:	Variable—first exposure 100 mA; second exposure 50 mA; third exposure 200 mA.

Comment. The student should note that there is not a marked density difference between exposures 1 and 2, but a considerable difference is seen between exposures 2 and 3. As the mA is increased, greater current is applied to the cathode filament. This results in greater film blackening.

Experiment 2. The relationship of exposure time and film density.

Purpose. To show the effect of the exposure time on film density, all other factors being held constant.

Theory. The amount of time allowed for an exposure is directly related to film density.

Procedure. Make 3 exposures using the following factors:

Absorber:	Hand phantom
Film:	Medium-speed 24 × 30 cm (10 × 12 in.)
Screens:	Medium-speed calcium tungstate
Grid:	None
Distance:	100 cm (40 in.)
kV:	50
mA:	100
Time:	Variable—first exposure 100 ms ($\frac{1}{10}$ sec); second exposure 50 ms ($\frac{1}{20}$ sec); third exposure 200 ms ($\frac{2}{10}$ sec).

Comment. The student should note that the film density of the first exposure is darker than that of the second exposure and lighter than that of the third exposure. If the equipment being used is properly calibrated, the film densities should be identical to those from Experiment 1. In Experiment 1, tube current (mA) was changed with kV and time constant. In this experiment, kV and mA are constant with exposure time variable.

Experiment 3. The relationship of mAs and film density.

Purpose. To show relationship between mAs and film density.

Theory. Film density should remain constant regardless of mA and time selections, provided mAs is constant. Thus, mAs = mA × time. If mA is increased or decreased, then time must be decreased or increased proportionately; and vice versa.

Procedure. Make a series of 4 exposures using the following factors:

Absorber:	Hand phantom
Film:	Medium-speed
Screens:	Medium-speed calcium tungstate
Grid:	None
Distance:	100 cm (40 in.)
kV:	50
mA:	Variable
Time:	Variable

Variable factors:	*mA*	*Time*
First exposure	100	100 ms ($\frac{1}{10}$ sec)
Second exposure	50	200 ms ($\frac{2}{10}$ sec)
Third exposure	300	33 ms ($\frac{1}{30}$ sec)

Comment: If the equipment is properly calibrated, film density for all four exposures should be identical, since the product of mA × time in all four exposures equals 10 mAs. These exposures should also be identical to the first exposures in Experiments 1 and 2.

Experiment 4. The relationship of kilovoltage and density.

Purpose. To show the effect a change in kilovoltage will have on film density.

Theory. As kilovoltage is changed, there is a corresponding change in film density.

Procedure. Make a series of 4 exposures using the following factors:

Absorber:	Hand phantom
Film:	Any 24 × 30 cm (10 × 12 in.)
Screens:	Cardboard holder 24 × 30 cm (8 × 10 in.)
Grid:	None
Distance:	100 cm SID (40 in.)
Collimate:	To phantom
kV:	Variable factor
mA:	100
mAs:	40
Time:	400 ms ($\frac{4}{10}$ sec)
Variable factors:	
First exposure	50 kV
Second exposure	60 kV
Third exposure	70 kV
Fourth exposure	80 kV

Comment. The overall density of the radiographs increases with the kV. Successive increases in kilovoltage cause corresponding increases in density because of greater tissue penetration due to higher photon energy and higher intensity x-ray beam; thus, less radiation is absorbed by the tissues and more reaches the film. Note also that the density change from exposure 1 to exposure 2 is not the same as from exposure 3 to exposure 4. There is not a linear relationship between changes in kilovoltage and film density.

Experiment 5. The relationship of source/image-receptor distance and density.

Purpose. To demonstrate the relationship between SID and radiographic density.

Theory. Any change in SID influences the intensity of radiation. The intensity varies inversely as the square of the SID. Since changes in distance produce changes in intensity, radiographic density changes.

Procedure. Make a series of 3 lateral radiographs of a skull phantom using the following factors:

Film:	Any 24 × 30 cm (10 × 12 in.)
Screens:	Medium-speed calcium tungstate
Grid:	8 to 1, fine-line

Distance:	Variable factor
kV:	80
mA:	150
mAs:	15
Time:	100 ms ($\frac{1}{10}$ sec)
Variable factors:	
First exposure	91 cm SID (36 in.)
Second exposure	100 cm SID (40 in.)
Third exposure	120 cm SID (48 in.)
Fourth exposure	183 cm SID (72 in.)

Comment. As the SID increases, the radiographic density decreases; and vice versa.

Experiment 6. The relationship of time, source/image-receptor distance, and density.

Purpose. To demonstrate the interrelationship of time of exposure, SID, and radiographic density when all other factors are constant.

Theory. The time required for a given radiographic density is directly proportional to the square of the SID when all other factors are constant: $T_1/T_2 = D_1^2/D_2^2$.

Procedure. Make these postero-anterior radiographs of the hand phantom, employing the following factors:

Absorber:	Hand phantom
Film:	Any 24 × 30 cm (10 × 12 in.)
Screens:	Cardboard holder 24 × 30 cm (8 × 10 in.)
Grid:	None
Distance:	Variable factor
Collimate:	To phantom
kV:	50
mA:	50
mAs:	Variable factor
Time:	Variable factor

Variable factors:		*Time*	*mAs*
First exposure	76 cm SID (30 in.)	$\frac{1}{2}$ sec	25
Second exposure	107 cm SID (42 in.)	1 sec	50
Third exposure	152 cm SID (60 in.)	2 sec	100
Fourth exposure	183 cm SID (72 in.)	3 sec	150

Comment. The densities of these radiographs are approximately the same because suitable exposure-time compensation has been made for each SID change in accordance with the inverse square law. It is recommended that once an SID is established for a given projection, it should be considered a constant.

Experiment 7. The relationship of milliamperage, source/image-receptor distance, and density.

Purpose. To demonstrate the interrelation of milliamperage and SID with radiographic density.

Theory. When all other factors are constant, the milliamperage required for a given exposure is directly proportional to the square of the SID.

Procedure. Make a series of 3 postero-anterior exposures of the hand phantom, employing the following factors:

Absorber:	Hand phantom
Film:	Any 24 × 30 cm (10 × 12 in.)
Screens:	Cardboard holder 24 × 30 cm (8 × 10 in.)
Grid:	None
Distance:	Variable factor
kV:	50
mA:	Variable factor
mAs:	Variable factor
Time:	Variable factor

Variable factors:	mA	SID	Time	mAs
First exposure	10	91 cm (36 in.)	2.00 sec	20
Second exposure	30	100 cm (40 in.)	1.75 sec	52.5
Third exposure	100	122 cm (48 in.)	1.20 sec	120

If the exact mA station listed is not available, use any combination of mA and time to give correct mAs.

Comment. As the SID changes, the mA and/or mAs may be used to compensate for the normal loss in density. In routine radiography, however, mA should be established as a constant for a given projection whenever possible.

Experiment 8. The influence of mAs on density.

Theory. No practical amount of mAs will compensate for inadequate kilovoltage.

Procedure. Make a series of 8 radiographs in the postero-anterior projection.

Absorber:	Hand phantom
Film:	Any 24 × 30 cm (10 × 12 in.)
Screens:	Cardboard holder 24 × 30 cm (8 × 10 in.)
Grid:	None
Distance:	100 cm SID (40 in.)
kV:	30
mA:	Variable factor
mAs:	Variable factor
Time:	Variable factor

Note: Be sure to check the tube-rating chart before performing these experiments.

Variable factors:	mAs
First exposure	15
Second exposure	30
Third exposure	60
Fourth exposure	120
Fifth exposure	240
Sixth exposure	480
Seventh exposure	960
Eighth exposure	70, but use 50 kV

Comment. As more mAs is applied to the tube filament, the film becomes darker, but there is insufficient detail present. Only in radiograph 8, exposed for 70 mAs at 50 kV, do we have a practical amount of mAs and sufficient kilovoltage to penetrate the part.

Experiment 9. The relationship of kilovoltage, milliamperes-seconds, and density.

Purpose. To show that satisfactory overall density can be made comparable by using mAs to compensate for the density effect of changing kilovoltage, provided the radiation penetrates the structure.

Procedure. A series of 4 lateral radiographs should be made employing the following factors:

Absorber:	Skull phantom
Film:	Any 24 × 30 cm (10 × 12 in.)
Screens:	Medium-speed 24 × 30 cm (10 × 12 in.)
Grid:	8 to 1
Distance:	100 cm SID (40 in.)
kV:	Variable factor
mA:	Variable factor
mAs:	Variable factor
Time:	Variable factor

Variable factors:	kV	mAs
First exposure	65	60
Second exposure	75	30
Third exposure	85	15
Fourth exposure	95	7.5

Comment. The density of all exposures should be essentially the same. The process of determining conversion factors will be discussed in Chapter 7.

Experiment 10. Collimation.

Purpose. To show the effect of collimation on film density and detail.

Theory. As the x-ray beam is collimated to the field of interest, less scattered radiation is produced and less film blackening occurs.

Procedure. Make 3 exposures using the following techniques:

Absorber:	Skull phantom, lateral position
Film:	Medium-speed
Screens:	Medium-speed calcium tungstate
Grid:	None
kV:	70
mA:	100
Time:	100 ms ($\frac{1}{10}$ sec)
Collimation:	Variable
Variables:	*Collimation*
First exposure	None: collimator wide open
Second exposure	Collimate to cranium
Third exposure	Collimate to sella

Comment. Film density will decrease as the x-ray beam is collimated to the field of interest (sella). Note that although film density decreases, the bone trabecular pattern surrounding the sella is seen better.

Repeat the above experiment using (1) higher kilovoltages and (2) large absorber.

Experiment 11. Grid effects.

Purpose. To show the effect of grids in removing scattered/secondary radiation.

Theory. As the x-ray photon flux strikes an absorber, some of the x-ray photons penetrate completely through the absorber without a change in direction (remnant radiation), some photons are scattered in all directions (scattered radiation), and some photons interact with atoms producing secondary radiation. Secondary and scattered radiation produce a variable, non-informational fog, which also increases film blackening.

Procedure. Make 3 exposures using the following data;

Absorber:	Pelvic phantom
Film:	Medium-speed
Screens:	Medium-speed calcium tungstate
Grid:	8 : 1, stationary, fine-line, or as available
kV:	70
mA:	300
Time:	200 ms ($\frac{2}{10}$ sec)
SID:	To grid focal distance

1. Using a standard cassette without grid, make an exposure of the pelvic phantom, process the film, and reload the cassette.
2. Place the grid on top of the cassette and the phantom on top of the grid;

center the x-ray tube over the cassette and make another exposure, using the same technique factors.

3. Process the film and reload the cassette.

4. Repeat step 2, but increase the technique by 20 kV.

Comment. In the first exposure, scatter/secondary fog obscures detail. In the second exposure, fog has been reduced, but detail is still not adequate because of decreased density due to photon absorption by the grid. In the third exposure, good detail is present, since technique factors have been increased to offset grid absorption.

Experiment 12. Grid/tube angulation and alignment.

Purpose. To show effect of angulating or misaligning the x-ray across the axis of the grid.

Theory. If the x-ray beam is not central to the grid, grid cutoff will occur.

Procedure. Make the following exposures, using the technique indicated:

Absorber:	Skull phantom, lateral position
Film:	Medium-speed
Screens:	Medium-speed, calcium tungstate
Grid:	8 : 1, fine-line, focused grid or as available
kV:	90
mA:	300
Time:	200 ms (²⁄₁₀ sec)
SID:	Within focal range of grid

1. Center the x-ray beam over the center of the grid/cassette, with the phantom placed on the grid/cassette.

2. Move the x-ray tube 10 cm towards the side of the grid/cassette, make an exposure, and process the film.

3. Repeat the procedure, but angulate the x-ray tube (1) with and (2) across the axis.

4. Repeat, but move the x-ray tube out of the focal range.

Comments. Grid cutoff would be evidenced by decreased film density.

The following experiments are not designed to be used as research tools, but to give the student a basic understanding of the variables in radiologic technology.

Experiment 13. Comparison of x-ray films.

Theory. The exposure of a film/screen combination to radiation will give identical results if the products are identical.

Procedure.

Absorber:	Aluminum step wedge

Film:	Variable
Screens:	Medium-speed calcium tungstate
Grid:	None
kV:	70
mA:	100
Time:	100 ms ($\frac{1}{10}$ sec)

1. Make certain that the same cassette and screens are used for all exposures in this experiment. Technical factors (kV, mA, time) should not be changed between exposures. All exposed films should be processed in the same processor while maintaining constant temperature and replenishing rates.

2. Make an exposure of the aluminum step wedge and process the film.

3. Repeat the exposure, using the test film(s).

4. Develop and calculate sensitometric curves for each exposed film. Plot both curves on the same graph paper.

5. Repeat the procedure, using phantoms to correlate H & D (sensitometric) curves with practical studies.

Comment. This experiment allows one to grossly compare x-ray films.

Experiment 14. Split film technique for comparison of x-ray films.

Radiation output from x-ray equipment may vary as much as 25 percent from exposure to exposure. The amount of variation depends on many things: line voltage fluctuation, type and design of equipment, calibration of equipment, etc. To ensure that a test film receives an identical exposure to that film against which it is being compared, the split film technique is sometimes used.

Procedure.

Absorber:	Skull phantom or lucite step-wedge.
Cassette:	14 × 17 in.
Screens:	Medium-speed calcium tungstate.
kV:	70
mA:	100
Time:	100 ms ($\frac{1}{10}$ sec)
Film:	

Test procedure: Take a 7 × 17 in. film and place it in the cassette. Place a second 7 × 17 in. test film into the same cassette so that each film is adjacent to the other. Place the absorber on the cassette so that half is over one piece of film and the other half over the other piece of film. Make an exposure collimating the x-ray beam so that the absorber is fully exposed. Process both films at the same time in the same processor. Evaluate the films visually and sensitometrically.

Comment: If 7 × 17 in. films are not available, take a pair of scissors and cut a 14 × 17 in. film into two 7 × 17 in. films. The use of this procedure eliminates machine variation from the test procedure.

Experiment 15. Evaluation of screen speed and resolution.

Purpose. To demonstrate the effect screen speed has on (1) film blackening, and (2) film/screen resolution.

Theory. For calcium-tungstate screens, screen speed is increased by increasing the thickness of the phosphor. This increased thickness allows more photons to be absorbed and converted into light photons. Increasing the phosphor thickness also increases image spread, thus decreasing resolution.

Procedure. Technique factors:

Absorber:	Skull phantom, lateral; resolution target
Film:	Medium-speed
Screens:	Variable—1. Medium-speed calcium tungstate. 2. High-speed calcium tungstate
Grid:	None
kV:	70
mA:	300
Time:	100 ms ($\frac{1}{10}$ sec)

1. Make an exposure of the phantom and resolution target, using the medium-speed calcium tungstate screens. (Place the resolution target beside the phantom.) Process the film.
2. Using the same technique factors, make an exposure of the phantom and resolution target, but use the high-speed screens.
3. Decrease mAs 50 percent and repeat item 2.
4. Repeat the procedure, using available screens.

Comment. As screen speed increases, less radiation is required to obtain a specified density. The resolution, however, decreases.

Experiment 16. Film reciprocity failure.

Purpose. To demonstrate the effect exposure time has on film blackening.

Theory. Normally, any combination of milliamperage and time can be chosen for an exposure, as long as their product (mAs) is constant, provided that exposure equipment is properly calibrated. Deviation from this constant relationship is known as film reciprocity failure.

Procedure. Technique factors:

Absorber:	Aluminum step wedge
Film:	Variable
Screens:	Medium-speed or detail calcium tungstate
Grid:	None
kV:	70
mA:	25

Variable factors:

	Time		Distance (inches)
First exposure	2.0	sec	100
Second exposure	1.0	sec	70
Third exposure	.5	sec	50
Fourth exposure	.25	sec	35
Fifth exposure	.125	sec	25
Sixth exposure	.06	sec	17

7 TECHNICAL CONVERSIONS FOR RADIOGRAPHIC EXPOSURES

There are many occasions in radiography when one needs to change technical factors. One may have to increase kilovoltage to shorten exposure times or to better penetrate a part or to change contrast levels. Thus one must learn the basic principles of technique conversion. For example, suppose that optimum exposure techniques have been developed for a certain radiographic examination, but owing to the inability (voluntary or involuntary) of the patient being radiographed to cooperate, kilovoltage must be increased so as to allow for a shorter exposure time. As the level of the kV is raised, there has to be a concomitant decrease in mAs, which can be effected by decreasing mA, decreasing exposure time, or a combination of the two. The purpose of this chapter is to develop the basic concepts to follow when changing techniques for exposing radiographs.

RADIOGRAPHIC EXPOSURE

The word "exposure" has different meanings, depending on the context. For example, exposure may designate the radiographic conditions—kilovoltage (quality), milliamperage (quantity), and time—used for a certain technique. In somewhat different usage, exposure may indicate only the milliamperage and time (mAs) or source strength (quantity, for example of a cobalt unit) and time. Again, exposure sometimes designates exposure of the patient to radiation in terms of roentgens. *Absolute exposure* refers to a unit measure, such as milliroentgens per unit time, of the actual amount of radiation reaching a certain area of the film. *Relative exposure* is the amount of energy reaching a particular area of film and responsible for producing a particular density on the processed film in reference to a standard. Relative exposure is commonly used in the study of the sensitometric properties of x-ray films (see Chapter 2). As far as the x-ray film is concerned, theoretically it should make no difference what type of exposure is used, provided exposure factors are within reason.

THE PRIMARY EXPOSURE FACTORS

For any radiograph a "correct exposure" will require a precise combination of the four primary factors. These factors are:

  a. Kilovoltage (kV)
  b. Milliamperage (mA)
  c. Exposure time
  d. Source/image-receptor distance (focal-film distance)

Provided no other secondary factors—type of equipment, equipment calibration, screen/film combination, etc.—are changed, the radiographic density may be altered to a predetermined degree by varying any one of the four primary factors.

Three of these—milliamperage (mA), time *(t)*, and kilovoltage (kV)—are usually considered the factors controlled by the x-ray machine. The fourth factor is the distance from the focal spot of the x-ray tube to the image receptor which is called the source/image-receptor distance (SID), formerly known as the focal-film distance (FFD). The term milliampere-seconds (mAs) is descriptive of the milliamperage or amount of current applied to the filament of the x-ray tube for a given period of time. The radiographic density of a film varies directly with the mAs, provided the kilovoltage is sufficient to penetrate the part. The kilovoltage, spoken of as kilovolts-peak or kV, is the factor that controls penetration and contrast (see Chapter 5, Figure 5-8).

Kilovoltage affects the quality, and *to a lesser degree* the quantity of x-rays, and hence can be used as a variable to compensate for differences in part thickness. Under average conditions, variations in part thickness will require compensations of 2 kilovolts per centimeter thickness.

Milliamperage is a function of tube capacity and is important, when trying to avoid focal spot enlargement (blooming) and the selection of appropriate timing of the exposure. In earlier times it was quite impractical to use it as a variable, but it is a necessity today.

Time can be used as a variable with excellent results, especially where motion is a factor. In all moving parts there is a certain optimum exposure time beyond which motion becomes objectionable. This must be taken into consideration when determining exposure techniques.

Distance can be used as a variable, but is recommended only when generator limitations and motion make kV or mAs variations impractical. Changes in distance cause image magnification (or demagnification) with consequent variations in image detail, as has been discussed in preceding chapters. Moreover, calculations of technique factor changes for distance changes are time consuming; and focused grids, particularly of high ratio (12 : 1, 16 : 1), have very narrow focal ranges and thus limit distance as a variable.

For practicality and simplicity, it is better to keep three of the factors constant and to vary only the fourth to effect changes in radiographic density.

BASIC RULES FOR USING TECHNIQUE CONVERSIONS

There are a number of basic concepts the student must remember as technical factors are manipulated to obtain a specific examination. These rules are:

1. Selection of milliamperage must necessarily depend on the capacity of the x-ray generator and the x-ray tube loading, including the requirements of the focal spot dimensions. High milliamperage imposes the requirement of relatively large focal spot dimensions, which decrease the sharpness of detail.
2. Time of exposure should be reduced to a minimum to counteract the effects of motion. If the generator is calibrated correctly, for all practical considerations any combination of exposure times and milliamperage is permissible as long as mAs remains constant. The greater the milliamperage, the less will be the time of exposure required and vice versa. For example, an exposure of 100 mA at 100 ms is the same as 1000 mA at 10 ms. However, the trade-off for selecting high mA stations is decreased resolution.
3. Roughly, kilovoltage changes must be interpolated with respect to mAs values. Rules of thumb are available for kV technique changes, but these should be used only for kV ranges from 60 kV to 80 kV. For kV above 80 and below 60, charts on kV-mAs conversions (Tables 7-1 and 7-2) should be used. Similar changes in kilovoltage applied at the low (50 kV) and high ends (120 kV) of the diagnostic kilovoltage range do not produce identical changes in film density.

RELATION OF KILOVOLTAGE (kV), MILLIAMPERAGE-SECOND (mAs), AND DENSITY

Radiographic density varies greatly with the kilovoltage. In medical radiography, there is no precise mathematical method for determining kilovoltage-milliamperage-seconds density ratios. Such factors as the thickness and density of the body tissues to be examined, the characteristics of the x-ray apparatus, and whether the film is used with or without intensifying screens exert pertinent influences. Fairly close approximations between kV and other exposure factors have been of necessity established by empirical means—by trial and error. There are procedures that may be followed in estimating the kV required for a given density change or for determining the approximate change in mAs required to compensate for a density change in kV.

Tables 7-1 and 7-2 were developed empirically from radiographs of living subjects on screen-type film. By the use of mAs multiplying factors, approximate mAs values may be obtained for changes in the kV range of 50 to 100 in increments of 5. Since the film response to x-ray exposure when screens are used differs from the response when direct exposure is used, two tables of values are required. Table 7-1 is for screen exposures, and Table 7-2 for direct x-ray

Table 7-1. Screen exposures table: to find mAs when kV is changed

Old kV	mAs multiplying factor									
50	1.0	.58	.38	.27	.19	.14	.09	.07	.05	.045	.035
55	1.74	1.0	.66	.47	.33	.23	.17	.13	.1	.08	.065
60	2.63	1.58	1.0	.7	.5	.35	.25	.19	.155	.13	.1
65	3.7	1.4	1.3	1.0	.7	.5	.35	.27	.22	.18	.14
70	5.25	3.0	2.0	1.4	1.0	.7	.5	.38	.31	.25	.2
75	7.54	4.3	2.8	2.0	1.4	1.0	.7	.54	.44	.35	.29
80	10.7	6.03	4.0	2.8	2.0	1.4	1.0	.75	.6	.5	.4
85	14.0	8.0	5.25	3.7	2.6	1.9	1.33	1.0	.8	.66	.54
90	17.0	9.7	6.45	4.57	3.2	2.3	1.6	1.22	1.0	.8	.66
95	21.4	12.3	8.0	5.6	4.0	2.8	2.0	1.5	1.22	1.0	.8
100	26.0	15.0	10.0	6.92	5.0	3.47	2.5	1.85	1.5	1.22	1.0
New kV	50	55	60	65	70	75	80	85	90	95	100

To use this table, locate the original kV in the left-hand (vertical) column and the desired kV in the bottom (horizontal) row. Where the corresponding row and column cross, the multiplying factor will yield the new mAs. *For screen exposures only.*

Example: Assume that 85 kV was used to radiograph a lateral skull. The mAs was 15. It is desired to use 100 kV in order to lengthen the scale of contrast and to stop motion.

Solution: Locate 85 kV in the left-hand (vertical) column and 100 kV in the bottom (horizontal) row. In the square where the corresponding row and column cross, we find the mAs multiplying factor to be .54. Thus 15 × .54 = 8.10 mAs— the correct mAs at 100 kV. The 8.1 mAs is not ordinarily available on modern x-ray control panels, and the nearest practical value should therefore be used; in this example, one would use 10 mAs.

Table 7-2. Direct exposures table: to find mAs when kV is changed (regular film)

Old kv	mAs multiplying factor									
50	1.	.69	.5	.39	.3	.29	.19	.14	.12	.1	.09
55	1.45	1.	.73	.56	.43	.34	.27	.21	.18	.15	.13
60	2.	1.37	1.	.77	.6	.47	.37	.29	.25	.21	.19
65	2.58	1.77	1.29	1.	.77	.61	.48	.37	.32	.27	.24
70	3.33	2.29	1.66	1.29	1.	.79	.62	.48	.42	.35	.31
75	4.21	2.89	2.1	1.63	1.26	1.	.79	.6	.52	.44	.39
80	5.33	3.66	2.66	2.06	1.6	1.26	1.	.76	.66	.56	.5
85	6.95	4.78	3.48	2.69	2.	1.65	1.3	1.	.87	.74	.65
90	8.	5.5	4.	3.1	2.4	1.9	1.5	1.15	1.	.85	.75
95	9.41	6.47	4.71	3.64	2.8	2.23	1.76	1.35	1.17	1.	.88
100	10.66	7.33	5.33	4.13	3.2	2.53	2.	1.53	1.33	1.13	1.
New kV	50	55	60	65	70	75	80	85	90	95	100

Locate the original kV in the left-hand (vertical) column and the new kV in the bottom (horizontal) row. Where the corresponding row and column cross, the multiplying factor will yield the new mAs. *For direct exposures only.*

exposures. Some mAs values derived from these tables may require the use of a time value that is not within the practical scope of the exposure timer. It is then necessary to employ the nearest practical value. The resulting density difference is usually so small that it is often not recognizable. In most instances the derived mAs values are approximations, and, if necessary, slight alterations should be made for a given projection and body tissue. X-ray generators may differ in their output, which may vary with kilovoltage secondary calibration differences. *Tables, therefore, cannot be expected to apply accurately to all generators, but they can provide a close guide to an anticipated value.* Conversion factors depend upon where the starting point is in relation to kilovoltage, taking 70 kV as a starting point. The following approximate conversions may be helpful:

To decrease kilovoltage by
10 kV requires twice the amount of mAs.
15 kV requires three times the amount of mAs.
20 kV requires four to five times the amount of mAs.

To increase kilovoltage by
10 kV requires a reduction to one-half of the mAs.
15 kV requires a reduction to one-third of the mAs.
20 kV requires a reduction to one-fourth of the mAs.

The charts in Table 7-3 can be used as a reference for increasing the mAs from two to eight times and/or decrease the mAs from one-half to one-tenth. All of the technique conversion rules are for single-phase equipment. For three-phase equipment, cut mAs in half.

Table 7-3. Changes to be made in kV as mAs is changed

	Decrease in kV to be made when mAs is increased			
Initial kV	mAs 2×	mAs 3×	mAs 4×	mAs 8×
50	44	41	38	31
55	49	45	43	37
60	53	49	46	41
65	55	52	49	43
70	60	55	53	46
75	65	59	55	49
80	70	64	60	53
85	74	68	64	55
90	78	72	68	58
95	80	74	70	60
100	83	77	73	63
105	88	80	76	66
110	91	82	78	68
115	94	85	80	70
120	98	88	82	72
125	102	91	85	74
130	106	94	88	76

Table 7-3 cont.

Increase in kV to be made when mAs is decreased

Initial kV	mAs ½	mAs ⅓	mAs ¼	mAs ⅛
50	56	62	66	76
55	64	70	74	85
60	70	76	80	94
65	75	81	86	104
70	80	88	94	115
75	86	96	104	127
80	94	106	115	141
85	102	115	125	153
90	110	125	135	—
95	115	130	141	—
100	118	132	146	—

RELATION BETWEEN KILOVOLTAGE AND TIME

It is often necessary to estimate the kilovoltage for a given change in exposure time, or what approximate change in exposure time is necessary to compensate for a desired change in kilovoltage. While it is not possible to give an exact rule for all cases, Table 7-4 provides estimates of the corrections that should be applied to kilovoltage or exposure time when either is changed. Please note that these tables hold only within the specific kilovoltage range. It must be kept in mind that an increase in kilovoltage with the exposure adjusted to maintain the same density will produce lower contrast in the radiograph; and conversely, a decrease in kilovoltage will produce higher contrast.

KILOVOLTAGE—mAs RELATION (15 PERCENT RULE)

To reduce the exposure (mAs) to one-half at any level of kilovoltage, add 15 percent more kilovoltage (see Figure 7-1).

Example: A technique of 100 mAs and 70 kV produces an optimum radiograph, but we need to reduce the exposure time to one-half of our original. What change in kilovoltage is required in order to halve the mAs?

Solution: To reduce the exposure time to half, we must add 15% of 70 kV to 70 kV.
    15% = 0.15
    70 kV × 0.15 = 10.5 kV
    10.5 kV + 70 kV = 80.5, say 80 kV.
We then reduce the mAs to half, and the new technique is 50 mAs and 80 kV.

Example: A technique of 10 mAs and 50 kV must be changed so as to reduce the exposure time to 5 mAs.

Solution: 5 mAs is half of 10 mAs. To make the reduction we must increase the kV by 15%.

15% = 0.15
50 kV × 0.15 = 7.5 kV
7.5 kV + 50 kV = 57.5, say 58 kV.

Our technique will now be 58 kV and 5 mAs.

Example: Using a technique of 300 mAs and 60 kV, we wish to reduce the exposure to 75 mAs.

Solution: 75 mAs is ¼ of 300 mAs. We arrange to halve the mAs twice.

Step 1　15% = 0.15
60 kV × 0.15 = 9.0 kV
9 kV + 60 kV = 69 kV, for halving mAs to 150
Step 2　69 kV × 0.15 = 10.35 kV, say 10 kV
10 kV + 69 kV = 79 kV for halving 150 mAs
　　　　　　to 75 mAs

Our technique will now be 79 kV and 75 mAs.

Table 7-4.　Estimate factors for correcting kV with time

For kilovoltage range 55–75

To decrease exposure time	Increase kilovoltage	
	with screens	*without screens*
25%	4 kV	8 kV
50%	10 kV	20 kV
75%	20 kV	40 kV

Example: Suppose that with intensifying screens a kV of 70 and an exposure time of 1 sec have been employed, and it is desired to decrease the exposure time to 0.5 sec. What kilovoltage would be required?

According to the table, the decrease in exposure time is 50%, which necessitates an increase in kV of 10. Thus, 70 + 10 = 80 kV.

For kilovoltage range 65–85

To increase exposure time	Decrease kilovoltage	
	with screens	*without screens*
25%	3 kV	6 kV
50%	6 kV	12 kV
75%	8 kV	16 kV
100%	10 kV	20 kV

Example: Suppose that with intensifying screens a kV of 80 and an exposure time of 4 sec have been employed, and it is desired to decrease the kV to 70. What exposure time would be required?

The required decrease in kilovoltage is 10 kV, which necessitates a 75% increase in exposure time:

75% of 4 sec = 3 sec
4 sec + 3 sec = 7 sec

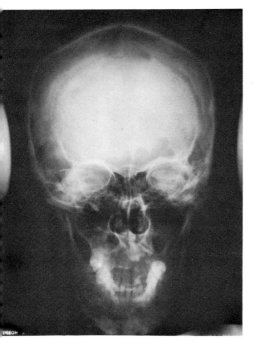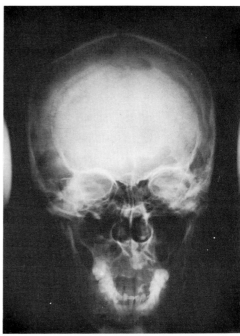

Figure 7-1. Comparative exposures of the skull made at 85 kV, 30 mAs (left) and 73 kV, 60 mAs (right). Increasing the exposure times by 100 percent requires a 15 percent decrease in kilovoltage.

KILOVOLTAGE AND IODINATED CONTRAST STUDIES

As a general rule, kV should not exceed 75 in performing iodinated contrast studies. This may be modified slightly by the use of different screens, grids, film, and three-phase generators. The reasons for this level of kV in these studies are the following:

(1) As kV increases, radiographic contrast decreases.
(2) The *linear attenuation coefficient* for fat and water (muscle) remains about the same as kV increases. (Linear attenuation coefficient is defined as the number (fraction) of x-ray photons which are removed from the x-ray beam per centimeter of absorber. The higher the density of absorber, the more the photons that will be removed or absorbed.)
(3) As kV increases, the linear attenuation coefficient for metals (i.e., iodinated contrast media, bone, barium sulfate suspensions) markedly decreases.
(4) In using iodinated contrast media, we try to maintain the maximum difference in linear attenuation coefficients between contrast media and soft tissues so as to better visualize the contrast media.

(5) Ideally, one should use about 65 kV for iodinated-related contrast studies. In actual practice exposure times may then be too long, or abdomens too large; so one would settle for about 70 to 80 kV, as the x-ray generators permit.

In many instances, the patient can be moved to a room having a larger x-ray generator, and we avoid increasing the kilovoltage. Under normal conditions, one would not want to use a high kilovoltage, which would give a rather long scale of contrast. This is particularly true in radiography of the biliary system, where low-density gallstones may be obscured by a long gray scale. However, one could choose a higher-speed screen film combination and thus use lower kilovoltage.

DISTANCE CONVERSIONS: THE INVERSE SQUARE LAW

X-ray intensities can be altered by moving the x-ray tube toward or away from the image receptor. Since x-rays obey the laws of light, a light bulb can be used to demonstrate this fact. With no other illumination in the room, move a single light bulb toward this printed page. You will find that the closer the light is to the page, the more brightly the page is illuminated; the farther the light is moved from the page, the less illumination of the page there will be. Exactly the same thing occurs with x-rays; the closer the source of radiation is to the image receptor, the higher the photon flux; and vice versa. There is a relationship between the distance of the x-ray source from the image receptor and the amount of radiation which will strike the image receptor, and hence the amount of film blackening which can be expected. This relationship is determined by the *inverse square law.*

We need to understand the term *inverse square* in its two parts—inverse and square.

Inverse is the opposite of direct. When we say, *"A varies directly as B,"* we mean "the more A, the *more B."* When we say, *"A varies inversely as B,"* we mean "the more A, the *less B."* Thus the greater the speed of a car, the more distance it covers in a given time; distance varies directly with speed. On the other hand, the greater the speed of a car, the *less* time it takes to cover a given distance; time varies *inversely* with speed. Again, if a weaver produces ten yards of cloth in a day, ten weavers will produce 100 yards in a day. There is a direct relationship between the number of weavers and the number of yards produced in a given time; the more weavers, the more yards. On the other hand, if it takes a weaver ten days to produce 100 yards, how long will it take ten weavers? The relation is an inverse one; the more weavers, the *less* time. Ten times the number of weavers will do the job in *one-tenth* the time: 10/1 is the inverse of 1/10.

In radiography too we find an inverse relation—between the illumination or

Then

$$mAs_2 = \frac{mAs_1 \times D_2{}^2}{D_1{}^2}$$

Put simply, if we want to keep the same contrast and double the distance, we will have to multiply the mAs by 4.

Example: A technique calls for 80 kV at 183 cm (72 in.) SID (source/image-receptor distance), and it is desired to decrease the SID to 91 cm (36 in.). The old technique calls for 10 mAs. What would the new mAs be at 91 cm?

Solution:

1. $\dfrac{mAs_1}{mAs_2} = \dfrac{D_1{}^2}{D_2{}^2}$

2. Crossmultiply: $mAs_2 \times D_1{}^2 = mAs_1 \times D_2{}^2$

3. Dividing both sides by $D_1{}^2$, we have

$$\frac{mAs_1 \times D_2{}^2}{D_1{}^2} = mAs_2$$

$$\frac{10 \times 91^2}{183^2} = 2.5 \text{ mAs}$$

We would then use 2.5 mAs, 80 kV at a 91-cm (36 in.) SID.

Figure 7-3 shows a chest radiograph made at 183 cm (72 in.). Using the above formula, the same subject was then radiographed at a 91 cm (36 in.) SID. The results are shown in Figure 7-4.

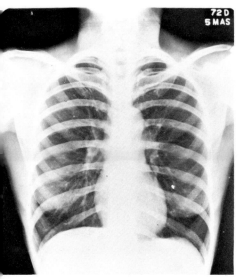

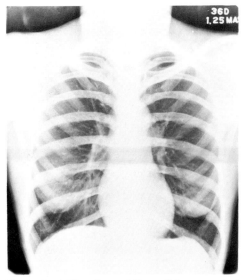

Figure 7-3. An example of mAs-SID conversion law. Compare with Figure 7-4. This radiograph was exposed at 183 cm (72 in.) and 5 mAs.

Figure 7-4. The chest of Figure 7-3, exposed at 91 cm (36 in.) and 1.25 mAs. In this case, distance was halved and the mAs was quartered. Note the obvious enlargement of the cardiac silhouette, caused by the short SID.

The mAs-distance formula (short method)

Mathematically

$$\frac{D_2{}^2}{D_1{}^2} = \left(\frac{D_2}{D_1}\right)^2$$

Hence, the mAs-distance conversion can be simplified by dividing the original distance into the new distance and squaring the result. Then, multiply the squared answer by the original mAs. This will give the new mAs to be used at the new distance.

Example:
>Old distance $=$ 183 cm (72 in.)
>New distance $=$ 91 cm (36 in.)
>Old mAs $\quad =$ 10
>New mAs $\quad =$?

Solution:

$$mAs_2 = mAs_2 \left(\frac{new\ distance}{old\ distance}\right)^2$$

Substituting known values, we find

$$mAs_2 = 10\left(\frac{91}{183}\right)^2 = 10\left(\frac{1}{2}\right)^2$$

$$= 10\left(\frac{1}{4}\right) = 2.5$$

Thus 2.5 mAs will be the new mAs to be used at the SID of 91 cm.

Example:
>Old distance $\quad =$ 91 cm (36 in.)
>New distance $=$ 183 cm (72 in.)
>Old mAs $\quad =$ 10
>New mAs $\quad =$?

Solution:

$$mAs_2 = mAs_1 \left(\frac{new\ distance}{old\ distance}\right)^2$$

Substituting known values, we have

$$mAs_2 = 10\left(\frac{183}{91}\right)^2 = 10(2)^2$$

$$= 40$$

Thus 40 mAs will be the new mAs to be used at 183 cm.

(Though these examples are simplified, division of any new distance by any old distance is a simple matter for the hand calculator.)

CORRECTION FACTORS FOR CHANGE IN SID

Table 7-5 lists the factors that apply for SIDs commonly employed in radiography. The initial SID is located in the left-hand vertical column and the desired distance is located in the horizontal column at the top of the table. The required mAs conversion factor will be found in the entry common to both columns.

Table 7-5.   mAs multiplying for changes in source/image-receptor distance

Old SID (inches)	New SID				
	36 in.	40 in.	48 in.	60 in.	72 in.
36	1.0	1.23	1.69	2.75	4.0
40	.81	1.0	1.44	2.25	3.24
48	.56	.69	1.0	1.56	2.25
60	.36	.44	.64	1.0	1.44
72	.25	.30	.44	.69	1.0

Example: A technique using a 40 in. SID is being used and it is desired to increase the SID to 60 in. What is the new mAs if 150 mAs is presently being employed?

Solution: We locate the old SID, 40 in., and follow its line to the right in the table to the column headed 60 in., the new SID. There we note 2.25, the multiplying factor. Thus 150 (old mAs) $\times$ 2.25 = 337.5 (new mAs sought).

Kilovoltage-distance conversion

When a variation from an established distance is found necessary, Table 7-6 will prove to be of value. However, the use of the mAs-distance relationship (mAs$_1$: mAs$_2$ :: $D_1^2 D_2^2$) will prove to be more accurate in maintaining radiographic density, and the contrast scale will not change, since kilovoltage remains the same. Table 7-6 should be used only when it is impractical to vary the mA and time values.

It is necessary to incorporate the 15 percent rule into the inverse square law if compensation for distance is going to be made with changes in kilovoltage. Compensating for distance changes with kilovoltage should be undertaken only as a last resort, primarily because of the changes in contrast it involves.

Table 7-6.   Changes in kilovoltage for changes in distance

From		To		Change in kV
cm	in.	cm	in.	(initial kV= 70)
101	40	122	48	add 5
101	40	152	60	add 13
101	40	183	72	add 18
183	72	152	60	subtract 5
183	72	122	48	subtract 12
183	72	91	36	subtract 18

Example: Let us assume a change from 101 cm (40 in.) to 183 cm (72 in.) and calculate kilovoltage changes required by use of the formula for mAs-distance relations and the 15 percent rule. Let us also assume an original kilovoltage of 80.

Step 1. By the formula for mAs/distance
$$\left(\frac{\text{New distance}}{\text{Old distance}}\right)^2 = \left(\frac{72 \text{ in.}}{40 \text{ in.}}\right)^2 = (1.8)^2 = 3.24$$
Thus, 3.24 times the radiation will be required at 72 in. as at 40 in.

Step 2. By the 15 percent rule, to double the radiation, add 15 percent of 80 kV to 80 kV.

80 × .15 = 12.0 kV

80 kV + 12 kV = 92 kV

Step 3. We still need a 1.24 increase in radiation (3.24 − 2.0 = 1.24). Since a 15 percent increase in kilovoltage will double radiation, let us set up a ratio to determine what percent increase will produce a 1.24 increase in radiation:

$$\frac{15\%}{2} = \frac{X}{1.24}$$

2X = 18.60%

X = 9.3%

Step 4.

92 kV + (92 kV × 9.3%) =

92 kV + 8.5 kV = 100.5, say 100 kV

Thus, 100 kV at 72 in. will produce the same density as 80 kV at 40 in. with mAs constant. Contrast, of course, will change.

TECHNIQUE CHANGES WHEN USING BEAM RESTRICTORS

Radiographic quality is visibly improved through the proper use of some beam-restricting devices. The device must restrict the primary beam to an area no larger than the field of interest.

Secondary/scatter radiation and fog increase with the amount and type of tissue struck by primary radiation, so by confining the field of primary radiation to only the useful area, secondary radiation emanating from tissue is reduced. Secondary-radiation fog affects visibility of image detail, radiographic contrast, and density. It reduces contrast by a graying effect. As an unpredictable supplemental density added to proper density, it makes density uncontrollable.

The amount of exposure compensation to be used for collimation is governed by

1. The proportionate decrease in the volume of the part being irradiated; this is governed by
   a. Thickness of the part
   b. The area of the collimation
   c. SID
2. The structure of the part
3. The kilovoltage

When the size of the x-ray beam is limited by a cone and/or a collimator, less tissue is irradiated and less secondary radiation is emitted from the part to fog the film. In general, when switching from a non-cone to a cone technique add 3 to 5 kV in order to compensate for loss in density due to the absorption of some of the primary radiation. Loss of density due to collimation is taken into consideration by technique charts.

Since x-ray film is either square or rectangular, any beam restrictor which permits a circular field of radiation large enough to cover an entire film contributes little if anything to minimize the effect of secondary-radiation fog. That is one reason why the once familiar flare cone has become obsolete. An extension cylinder, used alone or in conjunction with an adjustable diaphragm collimator, is a useful beam restrictor. If used properly it can be extended to confine a circular field of x-radiation inside the limits of even small-size films.

Conversion factors for cylinder cone

Extension cylinder collapsed—increase mAs 40 percent or 5 kV
Extension cylinder extended—increase mAs 60 percent or 10 kV

Conversion factors for rectangular collimators, field size (based on a 14 × 17 in. field

To convert to a 10 × 12 in. field—increase mAs 40 percent
To convert to an 8 × 10 in. field—increase mAs 60 percent

One can calculate the radiation required for changes in collimation by the inverse square law, which is explained above in this chapter.

GRID CONVERSION TECHNIQUES

Briefly, a grid is described in terms of grid ratio, lines per inch, focal range, and interspace material. Additionally, a grid is described in terms of the weight per square inch; this description, though it is perhaps the most accurate indicator of efficacy, is not in general use. Grid ratio is defined as the ratio of the height of the lead strip to the width of the interspace. Lines per inch relates to the number of lead strips per unit width. In radiography there are a number of grids with different lines-per-inch numbers available—e.g., the 60, the 80, and the 103. The radiographer should make certain that when grid conversion factors are calculated, the lines-per-inch number is taken into consideration. For example, technique factor changes required for converting from an 8 : 1/80 line to a 10 : 1/80 line are not the same as those for converting from an 8 : 1/80 line to a 10 : 1/103 line; an 8 : 1/80 line is radiographically equivalent to a 10 : 1/103 line. One should also note that the technique conversion factors for converting from a grid with an inorganic space material to an organic space material are not the same. Grid absorption is also affected by beam quality, which in turn is related to beam filtration and type of generator (three-phase, single-phase, and constant potential).

Thus it is difficult to commit to memory technique conversion factors for all the various and sundry available grids. Rules of thumb for grid conversions are just that—generalized rules to each of which there are as many exceptions as there are times in which the rule will work.

With few exceptions, a stationary grid or a Bucky mechanism should be used for radiography of all anatomical parts exceeding 12 cm in thickness. Use of a grid in radiography of the heart and lungs, and parts such as the knee, shoulder, and cervical vertebrae, is a matter of preference.

In pediatric radiography, certain body regions such as the skull, abdomen, pelvis, and vertebral column, even though they measure less than 12 cm in thickness, set up a considerable amount of secondary radiation when struck by primary radiation. Effective control of the fogging effect can be accomplished through use of a grid.

When converting from a non-grid technique to a grid technique, compensation must be made for the grid by an increase in either kV or mAs or both. The higher the grid ratio, the greater the absorption of primary and secondary radiation. Remember also that grids of identical ratio will not necessarily absorb the same amount of radiation, the amount of radiation absorbed being also a function of a strip density. Thicker parts and anatomical areas are affected greatly by secondary radiation and require a proportional increase in exposure factors.

For grid conversion techniques, it is preferable to correct the exposure factors by adding kilovoltage rather than milliamperage. This method of compensating for the grid assures adequate penetration and does not result in a great difference in radiographic contrast between non-grid and grid radiographs. Twenty kilovolts will provide adequate compensation for most modern high-ratio grids. There is an unnecessary exposure of the patient to radiation if compensation is made only in mAs.

When a grid is removed from a procedure, exposure technique factors must be decreased. In this instance, the compensation is preferably made by reducing mAs rather than kV if at all possible and within the limits of desired contrast levels. Tables 7-7 and 7-8 illustrate grid conversion techniques. Again it should be noted, these grid conversions will not work in all instances.

Table 7-7. Grid conversion factors in mAs[a]

From non-grid to:	mAs
5:1	Use 1½ times the mAs
6:1	Use 2 times the mAs
8:1	Use 3 times the mAs
12:1	Use 3½ times the mAs
16:1	Use 4 times the mAs

[a] The conversion factors are for an 80-line, organic interspaced grid. These figures are included only for reference purposes; technique changes for grids should be done with kilovoltage rather than with milliamperes-seconds.

Table 7-8. Grid conversion factors in kV

From non-grid to:	kV
5:1	Add 8 kV
6:1	Add 12 kV
8:1	Add 20 kV
12:1	Add 23 kV
16:1 [a]	Add 25 kV

[a]To change from an 8:1 grid to a 16:1 grid, add 6 kV or increase mAs by 30%. The correction factors are for an 80-line, organic interspaced grid.

THE PHOTOGRAPHIC EFFECT

To review, a combination of distance, time, kilovoltage, and milliamperage is required if a prescribed quantity and quality of radiation is to be presented to the image receptor (the film/screen combination in this case). The radiation in turn strikes a screen which absorbs, converts and emits a certain portion of light photons in proportion to the incident radiation. The x-ray film is primarily responsive to the light photons emitted from the screen. Changes in the quality and quantity of radiation therefore cause corresponding changes in the film response to the incident radiation. This response of the film to radiation and subsequent processing is called the *photographic effect.* For our purposes we will define photographic effect as primarily a change in film density with secondary changes in contrast. Since the aim of good radiography is to produce a radiograph with optimal density and contrast, it is our purpose to show how the four primary factors may be manipulated so as to maintain a constant film density.

If other factors remain constant, the intensity of an x-ray exposure, as measured by its photographic effect, varies directly as the milliamperage through the tube and directly as the time of exposure. If the voltage and distance are kept constant, the photographic effect of x-ray exposures varies directly as the number of milliamperage-seconds used. If the milliamperage is kept constant, the photographic effect varies directly as the variations in the time of the exposure; if the time is kept constant, the photographic effect varies directly as the variations in milliamperage. Thus, if an exposure is made with any milliamperage for a certain number of seconds, doubling the time will double the photographic effect; halving the time will halve the effect.

The photographic effect also depends on where along the film characteristic curve the exposure occurs. With an exposure occurring on the straight-line portion of the curve, the following rules hold:

1. Photographic effect (density) is directly proportional to the milliamperage applied to the x-ray tube filament. (PE $\propto$ mA)

2. Photographic effect (density) is directly proportional to the time (T) of the exposure. (PE ∝ T)
3. Photographic effect (density and contrast) is approximately proportional to the kilovoltage squared. (PE ∝ kV^2)
4. Photographic effect is inversely proportional to the square of the SID.

$$\left(PE \propto \frac{1}{SID^2}\right)$$

Mathematically, if photographic effect is related to the above four factors, then the four factors must be related to each other in the same fashion. This relationship can be written in the following manner:

$$PE = \frac{mA \times T \times kV^2}{SID^2} \text{ where}$$

PE = photographic effect
mA = milliamperage
T = time
kV = kilovoltage
SID = Source/image-receptor distance, in either inches or metric equivalents

Example: Calculate the photographic effect for the following exposure factors:
mA = 30
T = 1 sec
kV = 60
SID = 75 cm

Solution: Substituting in the above formula, we note

$$PE = \frac{30 \times 1 \times 60^2}{75^2} = \frac{30 \times 1 \times 3600}{5625} = \frac{108,000}{5,625} = 19.2$$

Thus, the photographic effect of the four factors is 19.2.

Example: Suppose that the above factors produced an acceptable radiograph, but for some reason the milliamperage must be changed, i.e.,
mA = 10
kV = 60
SID = 75 cm
What would be the new exposure time?

Solution: Since $PE = \frac{mA \times T \times kV^2}{SID^2}$, cross-multiply to solve for T, the unknown. We find

$$PE(SID)^2 = mA \times T \times kV^2$$

Dividing both sides of the equation by $(mA)(kV)^2$ so as to have T alone on one side of the equation, we find

$$\frac{PE(SID)^2}{mA(kV)^2} = \frac{\cancel{mA} \times T \times \cancel{kV^2}}{\cancel{mA}(\cancel{kV})^2}$$

$$\frac{PE(SID)^2}{mA(kV)^2} = T$$

or

$$T = \frac{PE(SID)^2}{mA(kV)^2}$$

Substituting the knowns in the formula, we have

$$T = \frac{19.20\ (75)^2}{10\ (60)^2} = \frac{19.20\ (5625)}{10\ (3600)} = \frac{108,000}{36,000} = 3$$

Hence the new time for the exposure is three seconds.

The above formula can be used to calculate any of the four factors as long as there remains only one unknown.

RELATION BETWEEN TIME AND DISTANCE

If two exposures are identical, then the photographic effect (PE_1) of one is equal to the photographic effect (PE_2) of the other. Therefore:

$$\frac{mA_1 \times T_1 \times KV_1^2}{SID_1^2} = \frac{mA_2 \times T_2 \times kV_2^2}{SID_2^2}$$

If milliamperage and kilovoltage remain constant, then $mA_1 = mA_2$, and $kV_1^2 = kV_2^2$, thus:

$$\frac{\cancel{mA_1} \times T_1 \times \cancel{KV_1^2}}{SID_1^2} = \frac{\cancel{mA_2} \times T_2 \times \cancel{KV_2^2}}{SID_2^2}$$

Collecting terms, we find

$$\frac{T_1}{SID_1^2} = \frac{T_2}{SID_2^2}$$

or

$$T_1 : T_2 :: SID_1^2 : SID_2^2$$

Thus, *the exposure time for a given exposure is directly proportional to the square of the source/image-receptor distance.*

Example 1: Suppose that an exposure of 10 sec (T_1) and an SID_1 of 75 cm have been employed for a certain examination. It is desired to maintain kilovoltage and milliamperage constant, but to decrease the SID to 60 cm (SID_2). What exposure time (T_2) would be required to have the same photographic effect?

Solution: Since

$$\frac{T_1}{SID_1^2} = \frac{T_2}{SID_2^2}$$

$$(SID_1^2)\ (T_2) = T_1\ (SID_2)^2$$

$$T_2 = \frac{T_1\ (SID_2)^2}{(SID_1)^2}$$

Substituting in the formula,

$$T_2 = \frac{(10)\ (60)^2}{75^2} = \frac{36,000}{5,625} = 6.4\ \text{sec}$$

Cahoon's Formulating X-Ray Techniques

Example 2: Suppose that an exposure time of 2 sec (T_1) and an SID of 183 cm (SID_1) have been employed and it is desired to decrease the exposure time to 250 ms (T_2, .25 sec). What distance (SID_2) would be required?

Solution:

1. $\dfrac{T_1}{T_2} = \dfrac{SID_1{}^2}{SID_2{}^2}$

2. Solving for SID_2,

$$SID_2{}^2 = \frac{T_2\,SID_1{}^2}{T_1}$$

$$SID_2 = \sqrt{\frac{T_2\,(SID_1)^2}{T_1}}$$

3. Substituting our known values in the formula, we have

$$SID_2 = \sqrt{\frac{.25(183)^2}{2}}$$

$$SID_2 = \sqrt{\frac{.25(33,489)}{2}} = \sqrt{\frac{8,372}{2}}$$

$$SID_2 = \sqrt{4,186} = 64.7 = 25.5 \text{ in. (use a hand calculator)}$$

$$SID_2 = \sqrt{100 \times 41.86}$$

$$SID_2 = 10\sqrt{41.86}$$

4. Using a calculator, we find the square root of 42 to be 6.47, thus

$$SID_2 = 10\,(6.4)$$

$$SID_2 = 64 \text{ cm (25 in.)}$$

RELATION BETWEEN MILLIAMPERAGE AND DISTANCE

In our basic photographic effect formula, we note

$$PE = \frac{mA \times T \times kV^2}{SID^2}$$

To determine the relationship between milliamperage and distance with time and kilovoltage remaining constant, the following manipulations are performed:

$$PE_1 = PE_2$$

$$\frac{mA_1 \times T_1 \times kV_1{}^2}{SID_1{}^2} = \frac{mA_2 \times T_2 \times kV_2{}^2}{SID_2{}^2}$$

Using the same reasoning as illustrated in the preceding section, let $T_1 = T_2$ and $kV_1{}^2 = kV_2{}^2$; thus

$$\frac{mA_1}{SID_1{}^2} = \frac{mA_2}{SID_2{}^2} \text{ or}$$

$$mA_1 : mA_2 :: SID_1{}^2 : SID_2{}^2$$

The milliamperage (mA) is then directly proportional to the square of the source/image-receptor distance.

Example: Suppose that 50 mA (mA_1) and an SID of 9 cm (SID_1) have been employed and it is desired to increase the distance to 183 cm (SID_2) in order to obtain sharper detail. What new mA would be required?

Solution:

1. $\dfrac{mA_1}{SID_1{}^2} = \dfrac{mA_2}{SID_2{}^2}$

2. Solving for mA_2

   $mA_2 = \dfrac{mA_1 (SID_2)^2}{(SID_1)^2}$

3. Substituting known values in the formula,

   $mA_2 = \dfrac{(50)(183)^2}{90^2} = \dfrac{50(33,489)}{8,100} = \dfrac{(1,674,450)}{8,100}$

   $mA_2 = 206$ milliamperes, say 200 mA.

RELATION BETWEEN MILLIAMPERAGE AND TIME

To derive this association, the photographic effect formula is used again. In this case allow kilovoltage and distance to remain constant, so that $(SID_1)^2 = (SID_2)^2$, $(kV_1)^2 = (kV_2)^2$, and $PE_1 = PE_2$, thus,

$$\frac{mA_1 \times T_1 \times kV_1{}^2}{\cancel{SID_1{}^2}} = \frac{mA_2 \times T_2 \times kV_1{}^2}{\cancel{SID_1{}^2}}$$

Hence $mA_1 \times T_1 = mA_2 \times T_2$.

Dividing both sides by a constant $(mA_2)(T_1)$, we have

$$\frac{mA_1 \times \cancel{T_1}}{(mA_2)(\cancel{T_1})} = \frac{\cancel{mA_2} \times T_2}{(\cancel{mA_2})(T_1)}$$

$$\frac{mA_1}{mA_2} = \frac{T_2}{T_1}$$

Thus, the milliamperage (mA) required for a given exposure is inversely proportional to the time (T).

Example 1: Suppose that 50 mA (mA_1) and an exposure time of 2 sec (T_1) have been employed and it is desired to increase the milliamperage to 100 (mA_2). What exposure time (T_2) would be required?

Solution:

1. $\dfrac{mA_1}{mA_2} = \dfrac{T_2}{T_1}$

2. Cross-multiplying and solving for T_2,

   $mA_2 T_2 = mA_1 T_1$

   $T_2 = \dfrac{mA_1 T_1}{mA_2}$

3. Substituting known values we have,

   $T_2 = \dfrac{(50)(2)}{100} = \dfrac{100}{100} = 1$

The new exposure time would therefore be 1 sec.

Example 2: Suppose that 50 mA (mA_1) and an exposure time of 0.5 sec (T_1) have been employed and it is desired to decrease the exposure time to 50 ms (T_2). What milliamperage would be required?

Solution:

1. $\dfrac{mA_1}{mA_2} = \dfrac{T_2}{T_1}$

2. Cross-multiplying and solving for mA_2:

$$(mA_2)\,(T_2) = mA_1 T_1$$

$$mA_2 = \dfrac{mA_1 T_1}{T_2}$$

3. Substituting known values,

$$mA_2 = \dfrac{(50)\,(500\ ms)}{50\ ms} = \dfrac{25,000}{50} = 500\ mA$$

Comments

The relationship between mA and time is one of the most commonly used technique conversions in radiography. The conversion is often made when it is necessary to employ high milliamperage in order to obtain the advantages of very short exposure times such as are necessary with radiographing children and developing exposure techniques for rapid film changers. The milliampere-seconds factor, indicated by the symbol mAs, directly influences radiographic density when all other factors are kept constant. It is the product of the milliamperage (mA) and the duration of the exposure time in seconds (s). Either factor, mA or s, may be changed at will to perform the required radiographic exposure as long as the product—mAs—remains the same.

In standardized radiography, correct machine calibration is a prerequisite, since no matter what mA and time increment value we are using, so long as the product, mAs, remains the same, the density level should remain the same. This holds true for all direct x-ray exposures. As we have seen, intensifying screen exposures may show some loss in radiographic density due to reciprocity failure. While long-time reciprocity failure is common, short-time reciprocity is uncommon in practical x-ray exposure conditions. Reciprocity failure is related to the response of the film, not the screen. Non-linear responses to changes in kilovoltage are not a problem when calcium tungstate screens are used, but do become a problem for certain rare earth screens.

It is highly recommended that the student become adept at computing mAs values. For example, an extremity technique at 50 mA for 0.1 sec equals 5 mAs; however, if a patient for some reason has had trauma and cannot hold the extremity still, it would be far better to use 100 mA and 50 ms ($\frac{1}{20}$ sec). This would equal 5 mAs, and motion unsharpness would be better controlled. It would be a more skillful and scientific approach to good radiography.

It requires at least a 30 percent increase in mAs to produce a significant increase or decrease in density. To produce a diagnostically acceptable radiograph which requires either an increase or decrease in density, the 30 percent figure can be employed to affect density changes in a scientific, mathematical manner.

Example: Assuming an exposure of 15 mAs at 80 kV, and the radiograph slightly underexposed, multiply 15 by 130 percent. Now we use 19.5 mAs (or 20 mAs) to correct this condition.

When the kilovoltage and processing are constant, serious underexposure or overexposure should be corrected by either doubling or halving the initial mAs value to change density.

RELATION BETWEEN MILLIAMPERAGE, TIME, AND DISTANCE

In our photographic effect formula

$$PE = \frac{mA \times T \times kV^2}{SID^2}$$

we note that the photographic effect equals the milliamperage multiplied by the time of the exposure and the square of the kilovoltage divided by the square of the source/image-receptor distance. We noted also in the preceding section that milliamperage is inversely proportional to the time of the exposure: as milliamperage is increased, exposure time must be decreased. Thus, one can manipulate milliamperage and time as long as this product (mAs) is kept constant, notwithstanding availability of choices of mA stations, timing selections in the generator, and calibration of the generator.

In this illustration, consider the relationship between mA, time, and distance where kilovoltage is held constant, i.e., $(kV_1)^2 = (kV_2)^2$ and $PE_1 = PE_2$. In this case, let us also substitute a new term mAs to equal mA $\times$ T, thus,

$$\frac{(mAs_1) \times kV_1^2}{(SID_1)^2} = \frac{(mAs_2) \times (kV_1)^2}{(SID_2)^2}$$

$$\frac{mAs_1}{(SID_1)^2} = \frac{mAs_2}{(SID_2)^2}$$

or

$$mAs_1 : mAs_2 :: (SID_1)^2 : (SID_2)^2$$

Since the milliamperage (mA) and the time (T) required for a given exposure are both directly proportional to the square of the SID, the product of the two mAs is also directly proportional to the square of the SID.

Example: Suppose that 60 mAS (mAs_1) and an SID of 90 cm have been employed, and it is desired to increase the SID to 183 cm (SID_2). What mAs factor would be required?

Solution:

$$1. \quad \frac{mAs_1}{(SID_1)^2} = \frac{mAs_2}{(SID_2)^2}$$

2. Cross-multiplying and solving for mAs_2, we have

$$mAs_2 = \frac{mAs_1 \times (SID_2)^2}{(SID_1)^2}$$

3. Substituting known values,

$$mAs_2 = \frac{(60)\ (183)^2}{90^2} = \frac{60\ (33,489)}{8,100}$$

$$= \frac{2,009340}{8,100} = 248\ mAs$$

Comment:

In the above example as in earlier examples in this chapter, we have solved conversion problems in a consistent mathematical manner. The use of the metric system, where 1 inch equals 2.54 centimeters, involves the use of larger numbers. To simplify the mathematics, consider the following:

$$\frac{X^2}{Y^2} = \left(\frac{X}{Y}\right)^2 = \left(\frac{X}{Y}\right)\left(\frac{X}{Y}\right)$$

In the above example where distance is double (91.5 to 183 cm, or 36 to 72 in.), the mathematics can be greatly simplified by combining terms:

$$\frac{(SID_2)^2}{(SID_1)^2} = \left(\frac{SID_2}{SID_1}\right)^2$$

Substituting:

$$\left(\frac{183}{91.5}\right)^2 = (2)^2 = 4$$

Thus the mAs can be corrected by multiplying the original mAs by 4.

The interrelation of all SIDs may be similarly expressed by means of a conversion factor. Any change in mAs necessitated because of a change in the SID may then be calculated by multiplying by the proper factor the initial mAs value. Table 7-5 lists the factors that apply for SIDs commonly employed in radiography.

STEREORADIOGRAPHY

An ordinary photograph made with a camera shows perspective; the relative positions and sizes of the various objects shown in the photograph are readily apparent. This is not true of a radiograph, because it is a shadow picture of overlying parts having various opacities to the x-ray; it is a two-dimensional display of three-dimensional anatomy. The single radiograph does not necessarily show the depth in the structures.

Stereoradiography is a process in which three-dimensional anatomy is displayed in a three-dimensional format. It is relatively simple for all modern x-ray machines, and almost any region of the body may be stereoradiographed with little effort. However, two requirements must be kept in mind: namely, perfect positioning and complete immobilization. Stereoscopic vision is directly dependent on the definition in the radiographic images as well as the degree of stereoscopic shift and a moderately long SID for creating a greater depth of focus.

Stereoscopic films are made by exposing two films; the part of the body and the film are in exactly the same position, but with a different position of the x-ray tube for each exposure. In stereoscopic views of the chest, the x-ray film cassette must be changed without disturbing the part; this is provided for by use of a cassette changer. Moreover, for stereo films of the chest the patient must not breathe between exposures.

In exposing stereoscopic films, the distance and direction of the shift of the tube between the two exposures are both of considerable importance. Normally the distance of the tube is controlled by the construction of the stereoscope in which the films are to be examined; if we use a stereoscope with a 25-in. distance between the mirrors and the viewboxes, a 2½-in. stereo shift would be required. Generally speaking, the tube shift in stereoscopic radiography should be 10 percent of the SID for a 25-in. viewing distance, 9 percent for a 28-in., and 8.5 percent for a 30-in. viewing distance. An error of ±0.5 in. will probably have no effect on the stereo capability of the set of films.

The direction of the tube shift in exposing stereoscopic film is controlled by the direction of the predominating lines in the parts being radiographed. The tube shift should be, as nearly as possible, at right angles to the line. In the thorax the predominating lines are the borders of the ribs; therefore the tube shift should be along the vertebral column. In making films of the long bones, the shift should be across the bones; for bones of the skull it may be in either direction, depending on what structures are to be investigated. In exposing stereoscopic films, the parts to be radiographed are placed in position as for a normal single radiograph. For the first film from the center position, the tube is moved in the proper direction one-half of the distance of the shift. It is now in position for the exposure of the first film, and when this has been made, the tube should be moved past the center position an equal distance and the second film exposed. Table 7-9 illustrates tube shifts for common SIDs.

Stereo shift and grids

With a 5-to-1 or 6.5-to-1 Bucky grid, one may stereo-shift across the grid strips. Tube centering is not critical, and the SID can be changed through a reasonable range. As grid ratio increases, cross-shift stereoscopic exposures cannot be made. The difficulty is in matching the density of the two films of the stereo pair. Cross-table stereo-shifting should be abandoned when using an 8-to-1 grid. When a 16-to-1 grid is used, cross-shift stereoscopic radiography is impossible because of grid cutoff. The radiographic tube must be centered precisely over the midline of the table and SID must be fixed at the focal distance of the grid. (Whenever an underexposed radiograph results from a technique that has been giving consis-

Table 7-9. Stereoscopic tube shifts for common SIDs under practical viewing conditions

	For stereo viewing distance of		
	25 in. (10%)	28 in. (9%)	30 in. (8.5%)
SID (inches)	use tube shift (inches) of:	use tube shift (inches) of:	use tube shift (inches) of:
36	3½	3¼	3
42	4	3¾	3½
48	4¾	4¼	4
60	6	5½	5
72	7¼	6½	6

Any tube shift can be calculated by solving the equation:

$$\frac{\text{tube shift}}{\text{SID}} = \frac{2.5625 \text{ (interpupillary distance)}}{\text{viewing distance}}$$

An error of ± ½ in. is acceptable.

tent results with a 16-to-1 grid, always check the tube centering; the tube may not have been centered within the focal range of the grid ("off-focus"), and this may account for the underexposure. However, it is more common for the x-ray tube to be off center than off focus, and in this case one side of the film is blank.)

If cross-hatch grids are used, it is barely possible with a 5 : 1 or 6 : 1 grid to tilt the tube just enough to accomplish a stereo shift in either direction. Cross-hatch grids are normally used for high kilovoltage at a fixed distance and the central ray aligned at 90° to the grid surface.

Stereoradiography: accessory nasal sinuses

Caldwell projection. In the postero-anterior projections of the sinuses, it is important that the stereoscopic shift be made vertical and front to back and never crosswise, in order to prevent overshadowing of the anatomy of interest by the oblique projection of the adjacent structures. The normal Caldwell projection should throw the shadow of the petrous portion of the temporal bone in the upper portion of the maxillary sinuses.

Waters projection. When using the Waters projection for the maxillary sinuses, the shadow of the petrous portions of the temporal bone should be thrown below, or caudad to the antra. The tube shift should be made toward the top of the head (cephalad), so the antra will be free of superimposed structures.

Lateral projection. In the lateral projections, the tube shift may be horizontal or vertical depending on grid ratio.

Sphenoid, open-mouth projection. The tube shift should be horizontal (cephalad).

MAINTAINING SID AFTER THE TUBE IS ANGLED

Table 7-10 determines accurately where to place a tilted x-ray tube so that when the central ray is correct with the regional landmark, the anatomical region will be projected in the center of the film and the desired or original SID will be maintained.

If an x-ray tube is angled otherwise than 90° and the tube is left in its original position, the resultant radiograph will be underexposed.

Table 7-10. Data for tube angles used in radiography [a]

	Z 36 in.			Z 40 in.	
A	X	Y	A	X	Y
5°	2.8	35.8	5°	3.5	39.8
10°	6.2	35.5	10°	6.9	39.4
13°	8.0	35.0	13°	9.0	39.0
15°	9.3	34.8	15°	10.4	38.6
17°	10.3	34.4	17°	11.6	38.2
20°	12.6	33.8	20°	13.75	37.6
23°	14.0	33.0	23°	15.6	36.8
25°	15.4	32.6	25°	16.8	36.0
30°	18.0	31.2	30°	20.0	34.6
35°	20.6	29.5	35°	23.0	32.8
40°	23.2	27.6	40°	25.75	30.25
45°	25.4	25.4	45°	28.25	28.25

Z = SID; A = tube angle; X = tube shift; Y = new vertical distance. (Courtesy, Ralph Turrant, R.T.)

Example: AP view of the petrous bone, at a 36. in. SID; the tube has to be tilted 35°. What distance should you use at the 35° tilt? According to the chart on the 36 in. distance column we have: Z = 36 in. SID: A = 35° tilt; X = tube shift to have central ray over center of film; Y = new vertical distance, which will be 29.5 in.

[a] Modern radiographic films and screens are manufactured to allow for errors and adjustments of this nature. This table is included because it may still be useful for some technicians.

CAST RADIOGRAPHY

When using optimum kilovoltage techniques, measure the part and add 10 kV or double the mAs.

For lower-voltage techniques, measure the part and add 10 kV for a dry cast and 15 kV for a wet cast. It is better to alter kV than mAs. If, however, one prefers to use mAs, the correction factor would be to double the mAs for a dry cast and triple the mAs for a wet cast (Figure 7-5).

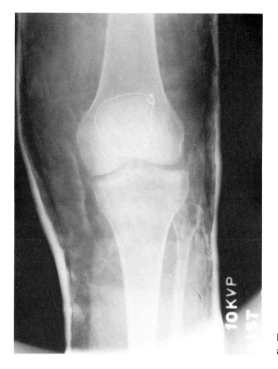

Figure 7-5. Radiograph of the knee in a cast, wi
a plus of 10 kV.

ADAPTING EXPOSURE TECHNIQUE FROM ONE HOSPITAL TO ANOTHER

From any given technique chart one may adapt the exposure factors to apply to any other machine.

Procedure. Choose a part from any group of parts of similar density. Make three exposures and divide the mAs from the chart being adapted into the mAs producing the most desired density with the new chart. The result obtained serves as the correction factor for all parts of similar density.

Example: A technique from Hospital A for the AP (antero-posterior) skull calls for 85 kV, 20 mAs, 100-cm SID, medium screens, 8-to-1 grid. It is desired to adapt this technique to Hospital B.

At Hospital B we make three exposures of the skull, employing all factors as used at Hospital A, except the mAs. Use the following three exposures:

    1. Below the original mAs;
    2. Equal to the original mAs;
    3. Above the original mAs.

We would use 15, 20, and 25 mAs. Process all three films at the same time. We now find that the film exposed at 25 mAs produces the desired density. This means that if we raise all the mAs values in the grid-screen group by an amount proportional to the difference between the 20 mAs of Hospital A and the equally satisfactory 25 mAs of Hospital B, we should find all the techniques of Hospital B satisfactory:

$25 : 20 = 1.25 : 1$, i.e.,

$25/20 = 1.25$

Thus we can adapt the techniques of Hospital A to Hospital B by multiplying all mAs values in the screen-grid group by 1.25. All other factors remain constant.

THE ANODE HEEL EFFECT

Unexplainable radiographic density differences which could not always be attributed to the incorrect use of exposure factors have annoyingly occurred in radiographs in many x-ray departments. These annoyances have been largely attributable to improper alignment of the tube to the part, but many are related to the anode heel effect. With today's high-powered generators and high-capacity x-ray tubes with small angle targets, it is important to recognize the anode heel effect. Evidence of the heel effect does not signify a tube fault, for it is an advantageous characteristic of all x-ray tubes and is an invaluable means toward producing balanced radiographic densities (Figure 7-6).

Various investigators have drawn attention to the angular distribution of radiation intensity from the anode face. Experiments have been described in which ionization measurements were made of the intensities emitted at various angles. The results were illuminating: a wide range of intensities that could have positive radiographic effect was demonstrated. But little more than academic interest has been awakened for applying this inherent characteristic of all x-ray tubes— the "heel effect"—in order to influence balanced radiographic densities in the image.

The heel effect is a variation in x-ray intensity output (depending upon the angle of x-ray emission from the focal spot) along the longitudinal tube axis and in relation to the long axis of a film. The intensity diminishes fairly rapidly from the central ray toward the anode side of the x-ray beam; on the cathode side of the beam, intensity increases slightly over that of the central ray. The heel effect is of no value at long SIDs. When average or short distances are used on large film areas, the effect is most pronounced, particularly where decided differences in tissue densities require balancing of radiographic densities to avoid over- and underexposures within the same image.

The approximate percentage of x-ray intensity emitted by a tube at various angles of emission may be determined directly from photometric measurements of the radiographic blackening of an x-ray film. For radiographic purposes, this procedure is adequate and gives an indication of the approximate percentage of quantity variation to be expected when the x-ray beam falls on specific areas of different sizes of x-ray films at various SIDs. Figure 7-7 graphically represents the mean values of radiographic density measurements obtained from many intensities emitted by various x-ray tubes. This diagram, drawn to scale, shows radii

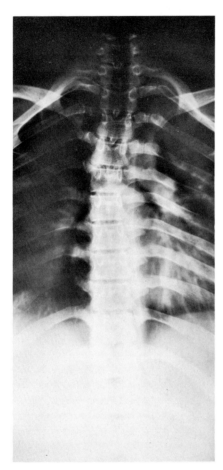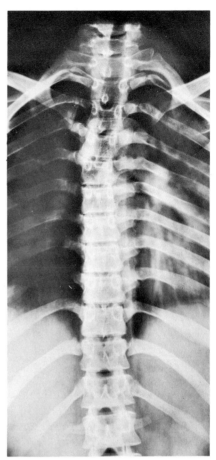

Figure 7-6. Antero-posterior screen-grid radiographs of thoracic vertebrae that demonstrate the density variation caused by the anode heel effect. Left, cathode portion of the x-ray beam directed cephalad. Right, cathode portion directed caudad (toward greater body thickness). Note more favorably balanced densities over entire image in the righthand radiograph.

emanating from the target face, drawn at 4° intervals from 0° to 40° and intercepting horizontal lines representing various SIDs. At the termination of the radii at the bottom of the chart, the mean density values in percentages are indicated. For convenience, the radiographic density caused by the central ray (CR) is figured as 100 percent. To the left or anode side of the central ray, the densities diminish in value, while those to the right or cathode side increase moderately and then decrease slightly.

The spaces between the vertical dotted lines beginning just below the diagram of the anode and terminating at the 72-in. SID line, when approximately paired, indicate the length of each size of film and the approximate location at various SIDs of the respective density values created by the intensities of radiation at various angles of x-ray emission. For example, on the 48-in. SID line the outermost pair of vertical lines representing the 36-in. length of film passes outside the limits of the x-ray beam. In order to make use of the entire range of intensities on this length of film, it would be necessary to employ an SID of 49 in. All intensity radii would intercept at this distance; consequently, if an exposure were made, the entire range of intensities would be expected to become evident, let us say, on a radiograph of an entire spine. Since the minimal intensity in the anode portion of the beam, approximately 31 percent, is emitted in advance to the angle indicated as 0°, it is obvious that the portion of the spine having least tissue density (the neck) should be exposed by this portion of the beam, and the heavier portion (the lumbar vertebrae) by the cathode portion of the beam. The above discussion holds true for a 20° anode; for smaller target angles used in many x-ray tubes today, this anode heel effect becomes even more critical.

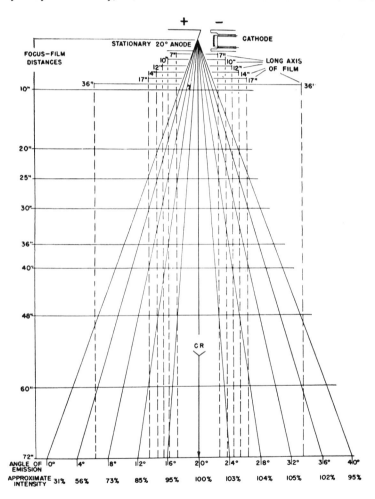

Figure 7-7. The anode heel effect for a 20° target angle. This change occurs over a much smaller area as the target angle steepens. (Courtesy, Eastman Kodak Co.)

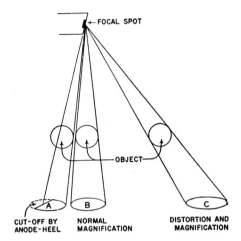

Figure 7-8. Limiting factors in short-focus radiography on large film sizes.

When radiographs on large films are made at relatively short SIDs, a knowledge of the distribution of the intensity in approximate percentages delivered to a particular size of film at a known SID is indispensable for correct alignment of the tube and part with the film at the most favorable SID. Radiographic densities cannot be balanced if only the intensity of the central ray is used (Figure 7-8).

Radiographs are often obtained in which a certain area of the image is definitely underexposed while another is overexposed, even though the factors seemed adequate for the tissue-density traversed by the central ray. This condition usually occurs when improper alignment of the tube with the part takes place. This lack of balance between radiographic densities is typical of the heel effect of the anode when it is improperly applied (Figure 7-6).

The general rule for utilizing the heel effect is this: *Align the long axis of the tube parallel with the long axis of the part to be examined, and direct the cathode portion of the beam toward the anatomic area of greatest tissue density.*

PEDIATRIC CONVERSION FACTORS

Technical conversion factors for children are as arbitrary as anything in radiology. The child's body tissues have greater water content than an adult's, and body structures are smaller; so technique factors must be adjusted. Only from long experience can one develop technique charts for pediatrics, and these must come primarily from trial and error.

Numerous texts list conversion factors for age groups. The only consistent thing is that there is no consistency, either in grouping by age or in technique conversion factors. One thing is certain: the child's body has more water content proportionally than an adult's, and more water means more scattering; thus, technique factors, principally kilovoltage, must be reduced.

Children can be classified into groups if one realizes that these groupings are purely arbitrary. The first group is from birth to about age 2 and is called infancy. The second group, called the preschool group, is from age 2 to age 5 or 6. The third group is the school group and ranges from ages 6 to 10 for girls and ages 6 to 12 for boys.

For the most part, children over 12 years (adolescents or teenagers) will require adult radiographic techniques.

As a rule of thumb, these starting points in the following technique conversions can be used:

Infancy	Birth to 2 years	Decrease mAs	75 percent
Preschool	2 to 6 years	Decrease mAs	50 percent
School age	6 to 12 years	Decrease mAs	25 percent
Teenage	More than 12 years	No change	

Other points to remember include these:

1. Use the shortest exposure time practical, to minimize motion unsharpness. Be careful that exposure times are not so short that grid motion is frozen, resulting in grid lines in the radiograph.
2. Use technique factors which allow optimum visualization of detail in relationship to the amount of radiation exposure. This also invariably means the use of fixed-kV techniques.
3. Gonadal shielding of some type is mandatory.

HIGHER-KILOVOLTAGE RADIOGRAPHY

The only limit to the kilovoltage in radiography is that imposed by contrast. The limiting kilovoltage is the one in which contrast has been reduced to the point that density differentiation or diagnosis becomes doubtful. All who have investigated the use of higher kilovoltages for radiography realize its contrast limitation.

The immediate acceptance of any technical procedure that alters to a marked degree the appearance of the radiograph is not to be expected. Everyone recognizes the need for maximum contrast in some types of work, but there are examinations in which increased penetration and increased exposure latitude are desirable and necessary. In Figure 7-9A, the lateral radiograph of the skull at 70 kV shows pathology in the occipital region. Figure 7-9B is the same skull at 60 kV, and Figure 7-9C at 100 kV. Note the changes in contrast and detail in the visualization of the pathological and anatomical structures. Generally, few advocate the use of higher kilovoltages (100 kV and above) for all examinations; but the higher kilovoltages have been widely accepted in studies of the gastrointestinal tract and colon, in the lateral view of the lumbosacral spine of obese patients, in pregnancy studies, and in various pathological studies. Langfeldt routinely uses

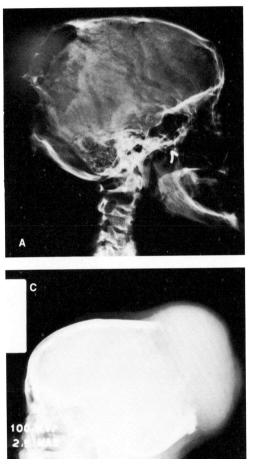

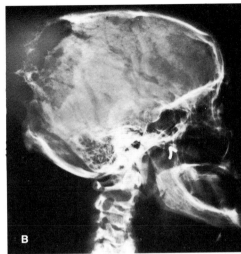

Figure 7-9. Effects of kilovoltage: **A.** A radiograph of lateral skull, with pathology, at 70 kV. **B.** The same skull, at 60 kV. **C.** The same skull, at 10 kV.

90 to 105 kV for demonstration of the auditory ossicles. The use of the higher kilovoltages is indicated because of better penetration of dense structures of the petrous bone and fluid, reduced exposure of the patient, and increased exposure latitude.

Some radiologists routinely use 150 kV for chest radiography with an 8-to-1 ratio grid. In some of the Scandinavian countries the kilovoltages have been increased to 150 kV, and the radiographic quality equals that of work done at a lower range. Exposure tables have also been worked out, based on the circumference of the patient's chest.

Radiography in the supervoltage range is of interest in selected cases because of the possibility of visualizing lesions obscured by bone, especially in the chest. Tuddenham and his associates have experimented with million-volt x-ray equipment for diagnostic purposes.

Radiographs made with the higher kilovoltages are satisfactory when the exposure factors have been correctly balanced. It is perfectly possible to produce good-quality, fog-free radiographs of the thicker parts with a higher kilovoltage. Good-quality radiographs employing different kilovoltages can be made as long as there are suitable compensations in mAs. However, the higher kilovoltages should be avoided in radiography of thin parts, i.e., the hand, or when using iodinated contrast material. Figure 7-10 shows examples of high-kV exposures.

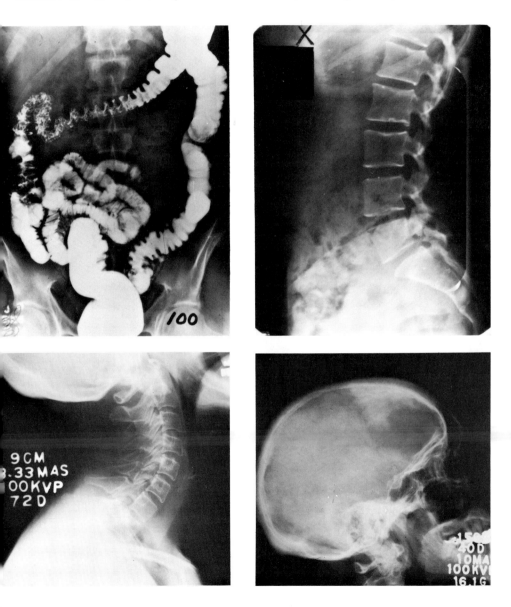

gure 7-10. Examples of high-kilovoltage technique.

TECHNIQUE CONVERSION TO HIGHER VOLTAGE RANGE

The graph of Figure 7-11 shows a conversion curve, based on visual densities obtained, of radiographs throughout the range of 40 to 130 kV. With the conversion curve, it is possible to determine the change in mAs for any desired kilovoltage, or the change in kilovoltage for any given change in mAs. Conversion would be made, of course, from any radiograph which produced the satisfactory degree of density in the normal voltage range. There are two curves on the chart—one with a solid line and one with a dotted line. The solid line represents constant density throughout the voltage range. However, a slight decrease in density can be tolerated if penetration is adequate, with a considerable gain in radiographic contrast; this is represented by the dotted line.

The equation for conversion is this:

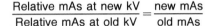

$$\frac{\text{Relative mAs at new kV}}{\text{Relative mAs at old kV}} = \frac{\text{new mAs}}{\text{old mAs}}$$

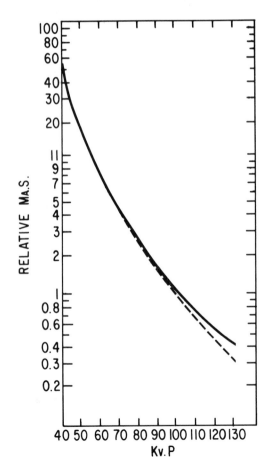

Figure 7-11. Conversion curve of density vs. kilo volts.

Problem 1: Suppose a lateral skull radiograph is made using the following factors: 20 mAs and 70 kV. It is desired to change to 5 mAs. Find the new kV.

Solution:

Step 1: Determine the relative mAs at starting kV (70). This, on the graph of Figure 7-11, is about 4.5.

Step 2: The relative value 4.5 is inserted into the equation so that the equation now reads;

$$\frac{\text{Relative mAs at new kV}}{4.5} = \frac{\text{new mAs}}{\text{old mAs}}$$

Step 3: The stated mAs factors in the problem are inserted into the equation and x is used for the unknown, so that our equation now reads:

$$\frac{x}{4.5} = \frac{5}{20}$$

Step 4: The problem is solved (ratio and proportion):

$20x = 22.5$

$x = 1.125$, relative mAs at the new kV

Step 5: This new relative mAs value (1.125) is located on the curve of Figure 7-11, and the new kV will be found directly below, on the horizontal axis of the graph—is in this example, 90 kV.

Problem 2: Suppose we have a technique using the factors 200 mAs and 70 kV; we wish to use 110 kV. What would the new mAs factor be?

Solution:

Step 1: Using the equation that precedes, in Problem 1, and the conversion curve of Figure 7-11, the relative mAs at the new kV will be 0.6, so the formula will read:

$$\frac{0.6}{\text{Relative mAs at old kV}} = \frac{\text{new mAs}}{\text{old mAs}}$$

Step 2: Relying on the conversion curve, the relative mAs at old kV will be 4.5, so the equation now reads:

$$\frac{0.6}{4.5} = \frac{\text{new mAs}}{\text{old mAs}}$$

Step 3: The stated mAs value of 200 is now inserted into the equation:

$$\frac{0.6}{4.5} = \frac{\text{new mAs}}{200}$$

Step 4: x is inserted to replace "new mAs," and the equation is solved:

$$\frac{0.6}{4.5} = \frac{x}{200}$$

$4.5x = 120$

$x = 26.6$, the new mAs

HIGH-KILOVOLTAGE AIR-GAP TECHNIQUE

It has been well established that the part to be radiographed should be as near the film as possible in order to obtain sharpness in the radiographic image. Definition or detail is good or bad according to the contrast in the radiographic image. Contrast depends partly on the amount of scattered radiation reaching

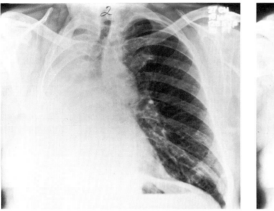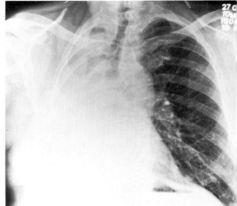

Figure 7-12. High-kilovoltage air-gap technique, with pathology: at left, a radiograph exposed at 125 kV; at right, a radiograph exposed at 150 kV.

the film, and a means must be provided for reducing the scatter to the film. Scattered radiation can be reduced by cones, diaphragms, and grids. Scattered radiation can also be reduced by an air gap such as is used in fractional-focus radiography. Watson has used an air-gap technique for radiography of the chest in large patients without a grid by increasing the part film distance by 6 in. and employing a 10-ft SID (Figure 7-12). Table 7-11 illustrates an air-gap technique for non-grid exposures at 125 kV.

Experiment. Photographic effect.

Purpose. To demonstrate the laws that if other factors remain constant, the intensity of an x-ray exposure, as measured by its photographic effect, varies directly as the milliamperes through the tube and directly as the time of the exposure. (The experiment requires hand processing.)

Procedure. Mark off with a pencil six equal areas on an 8 × 10 in. film in a cardboard holder, three squares on each side of a line that divides the film lengthwise into equal parts. Beginning at the upper left-hand corner and extending downward, number the areas on the left half from 1 to 3 and those on the right half from 4 to 6.

Use a small cone, 50 kV, and a 101-cm (40-in.) SID and the mAs given in the following table:

Area 1	150 mA	3 sec	150 mAs
Area 2	100 mA	1½ sec	150 mAs
Area 3	300 mA	½ sec	150 mAs
Area 4	100 mA	½ sec	50 mAs
Area 5	100 mA	1 sec	100 mAs
Area 6	100 mA	1½ sec	150 mAs

Table 7-11. Air-gap chest technique

Intensifying screens, non-grid, 120-in. SID, 6-in. air gap, kV 125

Anterior		Lateral	
cm range	mAs	cm range	mAs
16–17	1.6	22–23	3.3
18–19	2.5	24–25	5
20–21	3.3	26–27	6.6
22–23	5	28–29	8.3
24–25	6.6	30–31	10
26–27	7.5	32–33	13.3
28–29	8.3	34–35	16.6
		36–37	20
Oblique		38–39	25
cm range	mAs	40–41	30
22–23	6.6	Lordotic	
24–25	8.3		
26-27	10	cm range	mAs
28–29	13.3	18–19	6.6
30–31	16.6	20–21	8.3
32–33	20	22–23	10
34–35	25	24–25	13.3
		26–27	16.6

When the exposures have been completed, develop the film until the lightest area shows some blackening, disregarding the density of the other areas. Fix, wash, and dry the film.

Discussion. The finished film will show areas 1 to 3 of nearly the same density. Areas 4 to 6 will show a progressive increase in the density of the different areas. In areas 1 to 3 the milliamperes through the tube have been increased from 50 to 300, but the time has been correspondingly decreased. For each of these areas, a quantity of x-rays equal to 150 mAs has been used. For areas 4 to 6 the quantity of x-rays has been increased from 50 to 150 mAs. This graded increase in quantity should cause a definite and regular increase in the density of these areas.

This experiment shows that with voltage and distance constant, the photographic effect of x-ray exposures varies directly as the mAs used. If the milliamperage is kept constant, it varies directly as the variations in the time of the exposure; if the time is kept constant, the photographic effect varies directly as the variations in milliamperage. Thus, if an exposure is made with any milliamperage for a certain number of seconds, doubling the time will double the photographic effect, and halving the time will reduce the effect to one-half.

REVIEW QUESTIONS

1. Common primary exposure factors include all of the below EXCEPT
   a. Kilovoltage
   b. Milliamperage
   c. Exposure time
   d. Source-image receptor distance
   e. X-ray generator

2. The exposure factors controlled by the x-ray generator include
   a. Kilovoltage
   b. Milliamperage
   c. Exposure time
   d. Only A and B are correct
   e. A, B, and C are correct

3. The term SID refers to
   a. Source-item distance
   b. Source-image receptor distance
   c. New term for focal-film distance
   d. A and B are correct
   e. B and C are correct

4. Beam quality is primarily controlled by
   a. Kilovoltage
   b. Exposure time
   c. Milliamperage
   d. Focal-film distance
   e. Size of the focal spot

5. As a general rule, a change in thickness of a body part of one centimeter will require ____ kV compensation:
   a. 1
   b. 2
   c. 3
   d. 4
   e. 5

6. Selection of a large (600 mA) milliamperage station as compared to a smaller (300 mA) in doing a knee at 100-cm SID will
   a. Cause decreased resolution
   b. Cause decreased x-ray tube loading
   c. Cause shorting of x-ray tube
   d. Have no effect on detail
   e. Require increased kilovoltage

7. "Blooming" of the focal spot is caused by
   a. High filament current
   b. Kilovoltage applied too high
   c. Exposure time too short
   d. Incorrectly calibrated kilovoltage
   e. Incorrect selection of intensifying screens

8. To increase kilovoltage from 70 to 80 generally requires a change in mAs of

    a. Twice the mAs
    b. One-half the mAs
    c. One-fourth the mAs
    d. Three times the mAs
    e. One-third the mAs

9. To convert from single-phase to three-phase exposure factors, the mAs should be

    a. Doubled
    b. Halved
    c. Tripled
    d. No change
    e. Quartered

10. An increase of 10 in kilovoltage at the low end (50 kV) of the diagnostic x-ray range, in relation to a similar change at the high end (120 kV), will have what effect on film density?

    a. A change of 10 kV at the low end will produce a greater density change than at the high end
    b. Changes of kilovoltage in respect to density are linear across the kV range
    c. Expected film density change should be the same
    d. Film density is controlled by mAs, not kV.

11. The "15% rule" in radiography indicates that

    a. 15% of x-ray examinations must be repeated owing to incorrect exposure factors
    b. To reduce the exposure (mAs) to one-half at any level of kV, add 15% more kilovoltage
    c. To control motion unsharpness, exposure time must be decreased 15%
    d. Distance must be increased 15% to compensate for double the exposure time

12. The kilovoltage selected for iodinated contrast studies generally should not exceed

    a. 50   d. 75
    b. 55   e. 85
    c. 65

13. As distance is increased, radiation intensity

    a. Increases
    b. Decreases
    c. No change

14. The inverse square law indicates that

    a. The size of the x-ray beam becomes smaller as distance is increased
    b. Intensity of radiation varies inversely as the square of the source/image receptor distance
    c. Intensity of the radiation varies inversely as the square of the object/film distance
    d. Field coverage decreases as distance increases

15. Using the inverse square law, when SID is halved, the exposure time required to produce equal density is

    a. Doubled
    b. Quartered
    c. Tripled
    d. Unchanged
    e. Halved

16. Failure of the film reciprocity law indicates that

    a. The longer the exposure, the darker the film
    b. There is a nonlinear response of x-ray film from direct x-ray exposures
    c. Screen exposures using different combinations of mA and time, but keeping mAs constant, may not have identical densities
    d. The shorter the exposure time, the darker the film

17. On collimating to the area of interest, x-ray exposure must be

    a. Increased
    b. Decreased
    c. No change

18. The most accurate method of determining grid conversion factors is to determine

    a. Grid absorption
    b. Grid ratio
    c. Grid lines per inch
    d. The amount of lead in 1 square inch of the grid
    e. Type of interspace material

19. As a general rule, grids should be used in radiographing body parts

    a. Greater than 12 centimeters
    b. Less than 12 centimeters
    c. Greater than 20 centimeters
    d. Less than 20 centimeters
    e. Only with low kilovoltages

20. To prevent unnecessary x-ray exposure of the patient, grid technique changes should be limited primarily to

    a. Changes in kilovoltage
    b. Changes in milliamperage
    c. Changes in distances
    d. Changes in the time of the exposure

21. As a general rule, a 12 : 1 grid may be used up to

    a. 60 kV
    b. 80 kV
    c. 100 kV
    d. 120 kV
    e. 150 kV

For the following questions use this equation:

$$PE = \frac{mA \times T \times kV^2}{SID^2}$$

where,

    PE = photographic effect
    mA = milliamperage
    T = time
    kV = kilovoltage
    SID = source/image receptor distance

Given the technique factors listed in Column A: to produce the same photographic effect, what change would be required if the changes noted in Column B were made? Calculate the value of the photographic effect first.

22.   Column A   Column B

     PE = ____   PE = ____
     mA =   20   mA =   40
     kV =   50   kV =   50
     SID = 100   SID = 100
     T =   2.0 sec   T = ____

23. PE = ____   PE = ____
     mA = 100   mA = ____
     kV =   70   kV =   70
     SID = 100   SID = 100
     T =   1.0 sec   T =   0.5 sec

24. PE = ____   PE = ____
     mA = 1000   mA = ____
     kV =   70   kV =   70
     SID = 100   SID = 50
     T =   0.1 sec   T =   0.2 sec

25. PE = ____   PE = ____
     mA = 500   mA = 500
     kV =   70   kV =   70
     SID = 100   SID = ____
     T =   0.5 sec   T =   0.1 sec

26. As a general rule, to expose stereoscopic films, a tube shift of ____% of SID is required?

    a. 5   d. 25
    b. 10   e. 50
    c. 20

27. When performing dry-cast radiography, as a general rule kilovoltage should be increased

    a. 5 kV   d. 20 kV
    b. 10 kV   e. 25 kV
    c. 15 kV

28. When determining the effect of absorption of the x-ray beam by the anode, radiation intensity

    a. Is more at the anode side of the center
    b. Is less at the anode side
    c. Does not change

29. Anode heel effect becomes more pronounced as

    a. Target angle is reduced
    b. Target angle is increased
    c. No change

30. For an exposure of the pediatric preschool group, mAs should be

    a. Increased 50%
    b. Decreased 50%
    c. No change
    d. Increased 25%
    e. Decreased 25%

⑧ OTHER TECHNIQUE CONVERSION SYSTEMS

Understanding technique conversions is essential for the exposure of good-quality diagnostic radiographs, and there is no substitute for knowing the basics presented in Chapter 7. Unfortunately, the human being often makes errors in mathematical calculations, even those as simple as making change in a store. Errors in calculations of exposure factors almost invariably entail a repeat exposure which costs time, money, and an additional exposure of the patient to radiation.

Over the years a number of devices have been offered to simplify technique conversions; most of these have not survived the test of time, but some are valuable supplements to the basic knowledge of technique conversions. Three of these use charts; the other is electronic. Of the chart types, the Kodak KvP-MaS Computer, the Dupont Bit System, and the Siemens Point System stand out as illustrative examples.

THE KODAK KvP-MaS COMPUTER

The Kodak KvP-MaS Computer is a circular slide rule developed in the late fifties (Figure 8-1). The conversion scale divides anatomical groups into two categories.

THE DU PONT BIT SYSTEM

The Du Pont Bit System was developed by Robert J. Trinkle of E. I. du Pont Photo Products Department in the early sixties. It is widely used throughout the radiologic technology profession and has been translated into a number of languages. (It has also become part of computer programs to print technique charts.)

The Bit System is based on the assignment of relative units of exposure called "Bits" to the technical factors of radiography. The term Bit was chosen by Trinkle because "bit" implies something small; in this case a small unit of exposure.

KODAK KvP-MaS COMPUTER

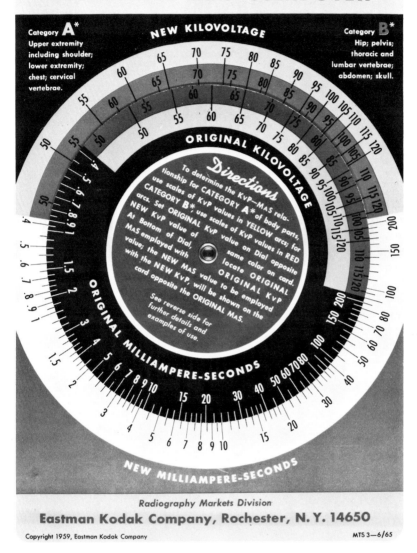

Radiography Markets Division
Eastman Kodak Company, Rochester, N. Y. 14650

MTS 3—6/65

Figure 8-1. The Kodak kilovoltage and milliampere-seconds computer. (Note the alternative abbreviation form for kilovoltage and milliamperes seconds.)

Bit is more commonly used as an abbreviation for binary digit-terminology used in computer storage and retrieval. We are also computing to obtain information.

The only calculations used in the Bit System are simple addition and subtraction, but not even these are needed with certain methods of use. The charts are designed so that one can simply measure distances on a scale for one exposure factor and then measure the same distance in another factor to find the technique change (Figure 8-2).

TABLE H		TABLE I	
kVp	**= Bits**	**mAs**	**= Bits**
153 =	11.4	640	= 22.0
147 =	11.2	550	= 21.8
141 =	11.0	485	= 21.6
135 =	10.8	420	= 21.4
130 =	10.6	365	= 21.2
125 =	10.4	320	= 21.0
120 =	10.2	280	= 20.8
115 =	10.0	240	= 20.6
110 =	9.8	210	= 20.4
106 =	9.6	182	= 20.2
102 =	9.4	160	= 20.0
98 =	9.2	140	= 19.8
94 =	9.0	120	= 19.6
91 =	8.8	105	= 19.4
88 =	8.6	92	= 19.2
85 =	8.4	80	= 19.0
82 =	8.2	70	= 18.8
80 =	8.0	60	= 18.6
78 =	7.8	52	= 18.4
76 =	7.6	45	= 18.2
74 =	7.4	40	= 18.0
72 =	7.2	35	= 17.8
70 =	7.0	30	= 17.6
68 =	6.8	26	= 17.4
66 =	6.6	23	= 17.2
64 =	6.4	20	= 17.0
62 =	6.2	17.5	= 16.8
60 =	6.0	15	= 16.6
58 =	5.8	13	= 16.4
56 =	5.6	11.2	= 16.2
55 =	5.4	10	= 16.0
54 =	5.2	8.7	= 15.8
53 =	5.0	7.5	= 15.6
51 =	4.8	6.6	= 15.4
50 =	4.6	5.7	= 15.2
49 =	4.4	5.0	= 15.0
47 =	4.2	4.4	= 14.8
46 =	4.0	3.75	= 14.6
45 =	3.8	3.33	= 14.4
44 =	3.6	2.90	= 14.2
43 =	3.4	2.50	= 14.0
42 =	3.2	2.16	= 13.8
41 =	3.0	1.80	= 13.6
39 =	2.8	1.66	= 13.4
38 =	2.6	1.45	= 13.2
37 =	2.4	1.25	= 13.0
35 =	2.2	1.08	= 12.8
34 =	2.0	.94	= 12.6
33 =	1.8	.83	= 12.4
31 =	1.6	.72	= 12.2
30 =	1.4	.63	= 12.0
29 =	1.2		
28 =	1.0		
27 =	.8		
26 =	.6		
25 =	.4		

(INCHES scales appear alongside both Table H and Table I)

Figure 8-2. Table H represents kV in terms of bits; Table I represents mAs. For example, to determine how much change in mAs would be required to change from 80 kV, 20 mAs to a 60-kV exposure measure from 80 on table H to 60 (1⅛ in.). On Table I measure *up* from 20 mAs 1⅛ in. The new mAs would be 80, to produce a radiograph of the same density.

The Bit System is compatible with all technique conversion systems and is in agreement with rules of thumb, percent relationships, and exposure factor relationships such as may be derived from the inverse square law and photographic effect formulas. The Bit System includes 19 different technical factors including estimates of skin dose; relationships of Bits to arithmetic speed and exposure factors; developing time, temperature, and activity; intensifying film-screen speeds; anatomical thicknesses; relationships of different kilovoltages and milliamperage-seconds to Bits.

There are rather distinct advantages to the Bit System. These advantages include:

1. Any technical factor which causes a change in film density can be theoretically calibrated into the Bit System.
2. Any two technical factors can be traded and balanced so that film density is practically unchanged.
3. It is a systematic approach to technique change which lends itself to research and improvement.

The Bit System is based on a calibration table for each technical factor. One cannot memorize these tables; therefore, a Bit Card has to be available for easy reference.

Before proceeding with examples of the Bit System, one should note the basic guidelines for use of the system:

1. Correct film density exists when kV Bits, mAs Bits, film-screen speed Bits, grid Bits, source-image receptor distance Bits, and anatomical thickness Bits are added together to equal a given Guide Bit Total.
2. Adding one Bit doubles exposure; subtracting one Bit halves exposure.

Table 8-1 illustrates a portion of the Bit System. The column on the left indicates kilovoltage and respective Bits, and the column on the right indicates mAs and its associated Bits. As an example of how the system is used, let us consider an exposure made at 66 kV, 20 mAs and we wish to make another exposure at 85 kV. What will the new mAs be?

Step 1: Obtain Bit total for original technique:

$$66 \text{ kVp} = 6.6 \text{ Bits}$$
$$20 \text{ mAs} = \underline{17.0} \text{ Bits}$$
$$\text{Total} = 23.6 \text{ Bits}$$

This total means that in terms of kVp and mAs this part needs 23.6 units of exposure in any combination.

Step 2: Subtract known new factor from total:

$$\text{Total} = 23.6 \text{ Bits}$$
$$\text{New 85 kVp} = \underline{-8.4} \text{ Bits}$$
$$15.2 \text{ Bits left for new mAs}$$

Step 3: From mAs table we note that 15.2 Bits correspond to 5.7 mAs which should be rounded up or down to the nearest generator setting.

Table 8-1. The Du Pont Bit System

kV-mAs conversions

kV = Bits	kV = Bits	mAs = Bits	mAs = Bits
153 = 11.4	58 = 5.8	640 = 22.0	17.5 = 16.8
147 = 11.2	56 = 5.6	550 = 21.8	15 = 16.6
141 = 11.0	55 = 5.4	485 = 21.6	13 = 16.4
135 = 10.8	54 = 5.2	420 = 21.4	11.2 = 16.2
130 = 10.6	53 = 5.0	365 = 21.2	10 = 16.0
125 = 10.4	51 = 4.8	320 = 21.0	8.7 = 15.8
120 = 10.2	50 = 4.6	280 = 20.8	7.5 = 15.6
115 = 10.0	49 = 4.4	240 = 20.6	6.6 = 15.4
110 = 9.8	47 = 4.2	210 = 20.4	5.7 = 15.2
106 = 9.6	46 = 4.0	182 = 20.2	5.0 = 15.0
102 = 9.4	45 = 3.8	160 = 20.0	4.4 = 14.8
98 = 9.2	44 = 3.6	140 = 19.8	3.75 = 14.6
94 = 9.0	43 = 3.4	120 = 19.6	3.33 = 14.4
91 = 8.8	42 = 3.2	105 = 19.4	2.90 = 14.2
88 = 8.6	41 = 3.0	92 = 19.2	2.50 = 14.0
85 = 8.4	39 = 2.8	80 = 19.0	2.16 = 13.8
82 = 8.2	38 = 2.6	70 = 18.8	1.80 = 13.6
80 = 8.0	37 = 2.4	60 = 18.6	1.66 = 13.4
78 = 7.8	35 = 2.2	52 = 18.4	1.45 = 13.2
76 = 7.6	34 = 2.0	45 = 18.2	1.25 = 13.0
74 = 7.4	33 = 1.8	40 = 18.0	1.08 = 12.8
72 = 7.2	31 = 1.6	35 = 17.8	.94 = 12.6
70 = 7.0	30 = 1.4	30 = 17.6	.83 = 12.4
68 = 6.8	29 = 1.2	26 = 17.4	.72 = 12.2
66 = 6.6	28 = 1.0	23 = 17.2	.63 = 12.0
64 = 6.4	27 = .8	20 = 17.0	
62 = 6.2	26 = .6		
60 = 6.0	25 = .4		

Table 8-2 illustrates how the Bit System can be used to determine mAs changes for changes in source-image receptor distances. From the table we note that 91 cm (36 inches) equals 7.0 Bits. If we were to increase the source-image receptor distance to 183 cm (72 inches), the Bits would *decrease by 2.0;* i.e., from 7.0 to 5.0 Bits. Referring back to Table 8-1 and assuming for this example that 30 mAs (17.6 Bits) was used for the exposure at 91 cm, we would have to *increase mAs by 2.0 Bits* to have a matched density exposure. Thus, we would require an increase from 17.6 Bits to 19.6 Bits. Again, referring back to the mAs chart we note that 19.6 Bits equals 120 mAs. From discussions in the previous chapter, we should remember that doubling the distance requires a four time increase in mAs. This is illustrated here since four times 30 equals 120 mAs.

The Bit System is available from your local Du Pont Technical Representative. He can also supply film-screen speeds and developer activities not shown on the Bit Card. Since this system is so commonly used, the tables of the 5th Edition are reproduced in Appendix A.

Table 8-2. SID conversions

Inches	= Bits	= Centimeters		Inches	= Bits	= Centimeters
6.7	= 12.0 =	17		39	= 6.8 =	99
7.3	= 11.8 =	18.5		41	= 6.6 =	104
7.7	= 11.6 =	19.5		44	= 6.4 =	112
8.3	= 11.4 =	21		48	= 6.2 =	122
9.0	= 11.2 =	22.5		51	= 6.0 =	130
9.5	= 11.0 =	24		55	= 5.8 =	140
10.0	= 10.8 =	25.5		58	= 5.6 =	147
10.6	= 10.6 =	27		63	= 5.4 =	160
11.4	= 10.4 =	29		67	= 5.2 =	170
12.2	= 10.2 =	31		72	= 5.0 =	183
13	= 10.0 =	33		77	= 4.8 =	196
14	= 9.8 =	35		83	= 4.6 =	211
15	= 9.6 =	38		88	= 4.4 =	224
16	= 9.4 =	40		95	= 4.2 =	241
17	= 9.2 =	43		102	= 4.0 =	259
18	= 9.0 =	46		109	= 3.8 =	277
19	= 8.8 =	48		116	= 3.6 =	295
21	= 8.6 =	53		125	= 3.4 =	318
22	= 8.4 =	56		136	= 3.2 =	345
24	= 8.2 =	61		144	= 3.0 =	366
25	= 8.0 =	64		154	= 2.8 =	391
27	= 7.8 =	69		165	= 2.6 =	419
29	= 7.6 =	74		176	= 2.4 =	447
31	= 7.4 =	79		180	= 2.2 =	457
33	= 7.2 =	84		203	= 2.0 =	516
36	= 7.0 =	91		218	= 1.8 =	554

THE SIEMENS POINT SYSTEM

The Siemens Point System is widely used throughout Europe and is fast becoming well known in this country. The Siemens System is based on the assignment of basic conversion factors called "points" to exposure techniques. Three points in terms of mAs either doubles or halves mAs; appropriate changes in kV do the same. To change from one set of exposure factors to another simply requires the addition or subtraction of an appropriate number of points.

The Point System consists of a series of exposure tables which list a number of individual anatomical areas. For each anatomical area an *initial exposure value point* (point) is listed. The initial exposure value presumes a non-grid exposure of an object of normal thickness, a 115-cm SID, medium-speed calcium tungstate screens, and medium-speed film. If any change in these basic factors occurs, one must apply correcting points for any addition or deletion. Other tables then indicate appropriate assigned points for certain kV, mAs, and distance conver-

Table 8-3. Siemens exposure tables

Part of body		Initial exposure value points	Normal object thickness cm	Distance cm	Screens	Grid Pb 8/40	Normal exposure value points	Recommended settings on two-pulse generators	
								kV	mAs
SKULL									
Skull survey	PA	30	19	115	Sapph.	with	34	73	100
Face bones	lat.	26	16	115	Sapph.	with	30	66	64
Neurocranium	lat.	27	16	115	Sapph.	with	31	70	64
Skull	axial	35	22	115	Sapph.	with	39	90	125
Os petrosa	sag.	33	17	115	Sapph.	with	37	77	160
Os petrosa, Stenvers		32	17	115	Sapph.	with	36	73	160
Sinuses	PA	32	22	115	Sapph.	with	36	73	160
Optical foramen, Rhese		29	17	115	Sapph	with	33	70	100
Mandible	lat.	24	11	115	Ruby	w.o.	27	60	50
CHEST									
Ribs 1–7		25	20	115	Sapph.	with	29	63	64
Ribs 8–12	PA	31	22	115	Sapph.	with	35	70	160
Sternum	PA	28	21	115	Sapph.	with	32	63	125
Sternum	lat.	29	30	115	Sapph.	with	33	66	125
Clavicle	PA	22	14	115	Sapph.	with	26	63	32
Scapula	AP	25	17	115	Sapph.	with	29	66	50
Scapula	lat.	28	14	115	Sapph.	with	32	63	125
Lungs	PA	19	21	150	Special	w.o.	18	60	9
Lungs	lat.	27	30	150	Special	w.o.	26	73	16
Heart	PA	21	21	200	Special	w.o.	24	85	5
Esophagus	obl.	30	28	70	Special	with	27	81	12
ABDOMEN									
Kidney, gallbladder	AP	29	19	115	Sapph.	with	33	66	125
Kidney, gallbladder	lat	35	27	115	Sapph	with	39	85	160

urinary bladder	axial	33	21	115	Sapph.	with	37	73	200
Stomach relief	PA	31	22	70	Special	with	28	85	12
Stomach filled	PA	32	22	70	or	with	29	85	16
Bulbus	PA	36	22	65	Diamond	with	31	120	8
Gast.-intest. surv.	elong.	31	22	115	Super	with	32	90	25
Obstetric pelvis	PA	32	32	115	"	with	38	90	100
Obstetric pelvis	lat.	39	28	115	"	with	40	90	160
Pelvimetry, Martius		38	29	115		with	39	90	125
SPINAL COLUMN									
Cervical vertebrae 1–3	oral	28	13	115	Sapph.	with	32	63	125
Cervical vertebrae 4–7	AP	26	13	115	Sapph.	with	30	63	80
Cervical vertebrae 1–7	lat.	23	12	150	Sapph.	w.o.	26	63	32
Cervical vertebrae 1–7	obl.	25	13	115	Sapph.	with	29	66	50
Upper thoracic vertebrae	AP	28	18	115	Sapph.	with	32	66	100
Lower thoracic vertebrae	AP	32	21	115	Sapph.	with	36	77	125
Thoracic vertebrae	lat.	32	30	115	Sapph.	with	36	73	160
Lumbar vertebrae 1–4	AP	30	19	115	Sapph.	with	34	73	100
Lumbar vertebrae 1–4	lat.	36	27	115	Sapph.	with	40	81	250
Lumbar vertebrae 1–4	obl.	33	22	115	Sapph.	with	37	81	125
Lumbar vertebrae 5	AP	33	22	115	Sapph.	with	37	81	125
Lumbar vertebrae 5	lat.	38	33	115	Sapph.	with	42	90	250
PELVIS									
Pelvis, hip	AP	30	20	115	Sapph.	with	34	73	100
Sacrum, coccyx	AP	31	19	115	Sapph.	with	35	77	100
Sacrum, coccyx	lat.	37	33	115	Sapph.	with	41	90	200
UPPER EXTREMITIES									
Shoulder	AP	22	11	100	Ruby	—	25	66	20
Shoulder	axial	23	11	100	Ruby	—	26	66	25
Upper arm	AP/lat.	19	8	100	Ruby	—	22	57	20
Elbow	AP	17	6	100	Ruby	—	20	57	12
Elbow	lat.	18	8	100	Ruby	—	21	57	16
Forearm	AP	17	6	100	R-Super	—	23	57	25
Forearm	lat.	18	2	100	R-Super	—	24	57	32

Table 8-3 cont.

Part of body		Initial exposure value points	Normal object thickness cm	Distance cm	Screens	Grid Pb 8/40	Normal exposure value points	Recommended settings on two-pulse generators kV	mAs
Wrist	d.v.	12	4	100	R-Super	—	18	50	16
Wrist	lat.	15	6	100	R-Super	—	21	55	20
Hand	d.v.	10	3	100	R-Super	—	16	50	10
Hand	lat./obl.	13	6	100	R-Super	—	19	55	12
Finger		8	2	100	R-Super	—	14	50	6
LOWER EXTREMITIES									
Neck of femur	lat.	32	33	100	Sapph.	—	32	66	100
Upper femur		27	13	115	Sapph.	—	31	70	64
Lower femur		24	12	115	Sapph.	—	28	66	40
Knee	AP	20	12	100	Ruby	—	23	60	20
Knee	lat.	19	10	100	Ruby	—	22	60	16
Patella, inf. sup.		20	12	100	R-Super	—	26	60	40
Patella	axial	19	7	100	Ruby	—	22	60	16
Tibia	AP	19	11	100	Ruby	—	22	60	16
Tibia	lat.	18	9	100	Ruby	—	21	60	12
Ankle	AP	20	9	100	Ruby	—	23	60	20
Ankle	lat.	18	7	100	Ruby	—	21	60	12
Heel bone	lat.	17	7	100	Ruby	—	20	60	10
Heel bone	axial	21	10	100	Ruby	—	24	60	25
Metatarsus	d.pl.	14	5	100	R-Super	—	20	60	10
Foot	lat.	16	7	100	R-Super	—	22	60	16
Metatarsus	obl.	15	6	100	R-Super	—	21	60	12
Toes		11	3	100	R-Super	—	17	55	8

sions. When points have been added or subtracted, the resulting point value is called *normal exposure value points.*

Table 8-3 lists the basic exposure points for a number of individual x-ray examinations and also indicates suggested exposure factors. Table 8-4 lists conversion factors for grids, SIDs, and screen speed. Table 8-5 lists the exposure data for mAs and kV. In any event, the exposure values listed in the table apply to the basic exposure factors noted in the paragraph above. When there are deviations from the normal part thickness, a point should be added to or subtracted from the normal exposure value for each centimeter of difference. However, in the

Table 8-4. Siemens exposure table

Target-film distance	= points	Grid		= points	
65 cm	− 5	Pb 8/40		+4	
75 cm	− 4				
85 cm	− 3	Pb 12/40		+6	
95 cm	− 2				
105 cm	− 1				
115 cm	0	*Intensifying*	*American (Du Pont)*	*= points at*	*= points at*
130 cm	+ 1	*screen*	*equivalent*	*50–90 kV*	*90–150 kV*
145 cm	+ 2	Siemens Sapphire	Medium	0	0
165 cm	+ 3	Diamond Super	Hi-plus	−3	−4
185 cm	+ 4	Siemens Ruby	2 × detail	+3	+3
210 cm	+ 5	Siemens Special		−3	−4
235 cm	+ 6	Ruby Super		+6	+6
260 cm	+ 7				
290 cm	+ 8				
325 cm	+ 9	*Miscellaneous*		*= points*	
360 cm	+10	Plaster cast		+4 to 5	
400 cm	+11	Narrow beam		+2	

Table 8-5. Siemens exposure table

mAs =	Points	mAs =	Points	mAs =	Points	kV =	Points	kV =	Points
0.1	−10	2	3	50	17	40	0	73	14
0.125	− 9	2.5	4	63	18	41	1	77	15
0.16	− 8	3.2	5	80	19	42	2	81	16
0.2	− 7	4	6	100	20	44	3	85	17
0.25	− 6	5	7	125	21	46	4	90	18
0.32	− 5	6.3	8	160	22	48	5	96	19
0.4	− 4	8	9	200	23	50	6	102	20
0.5	− 3	10	10	250	24	52	7	109	21
0.63	− 2	12.5	11	320	25	55	8	117	22
0.8	− 1	16	12	400	26	57	9	125	23
1	0	20	13	500	27	60	10	133	24
1.25	1	25	14	630	28	63	11	141	25
1.6	2	32	15	800	29	66	12	150	26
		40	16	1000	30	70	13		

case of chest radiography, one should subtract or add 1 point for each 1.5 cm change in thickness. If the density of the film made using the optimum normal exposure factors is less than optimum, the calibration of the equipment, film processing, and other factors should be checked.

To understand how this system is used, let us examine a postero-anterior skull exposure. Referring to Table 8-3 we note that the initial exposure value points for a PA skull are 30; the average centimeter thickness is 19 cm. Most skull exposures use a grid technique, and, from Table 8-4, 4 points must be added to the initial exposure value when adding an 8 : 1 grid. Thus a PA skull exposure requires 30 points for the initial exposure value plus 4 points for using an 8 : 1 grid, a total of 34 points. The recommended exposure is 73 kV, 100 mAs, single-phase radiographic equipment. Hence, the exposure factors are as follows:

Projection:	P.A. skull
Equipment:	Single-phase
Average cm thickness:	19 cm
Source/image distance:	115 cm
Screens:	Medium-speed calcium tungstate
Film:	Medium-speed
Exposure factors:	73 kV
	100 mAs

Initial exposure value points	30
Grid exposure 8 : 1	+ 4
Normal exposure value points	34

Technique conversions with Siemens point system

Suppose that in the above example it is desired to increase contrast, by lowering kilovoltage from 73 to 66. What new mAs is required?

Solution:

1. Initial exposure value   30
   Grid 8 : 1 exposure   + 4
   Normal exposure value points   34
2. From Table 8-4
   73 kv   = 14 points
   66 kV   = 12 points
   Difference   2 points
3. From Step 1 and from the preceding discussion we must maintain 34 points to have an optimum exposure. If we lower kV by 2 points, we must raise mAs by 2 points to maintain optimum film density. The question therefore is this: How much change in mAs is equivalent to 2 points?
4. From Table 8-5 (mAs)
   100 mAs = 20 points
   ? mAs = 22 points
5. From Table 8-5, 22 points is equivalent to 160 mAs. Hence, on lowering our kV from 73 to 66, mAs must be raised from 100 to 160.

As a comparison, the Kodak KvP-MaS Computer shows the change from 73 to 66 kV requires increasing the mAs from 100 to 160. The Du Pont Bit System shows a change from 73 to 66 kV requires an increase in mAs the equivalent of 0.7 bit. Since 100 mAs corresponds to approximately 19.3 bits, 20 bits (19.3 + 0.7) would be required for the exposure. Twenty bits is equivalent to 160 mAs. Thus, all three systems indicate the required change in technique factors to achieve the optimum exposure.

As long as the basic criteria are met (SID, normal part thickness, non-grid, and medium-speed screens and film, and the machine properly calibrated), any combination of kV and mAs points which adds up to the initial exposure value points will allow an optimum exposure, as long as the points for the kV and mAs do not exceed the initial exposure value. For example, if the initial exposure points for a PA skull exposure are 30, then any combination of mAs and kV points which add up to 30 can be used for the exposure *as long as the exposure is within the tube ratings and the contrast scale is desirable.* Thus for a PA skull exposure the following technique factors are possible:

mAs	pts	kV	pts
10	10	102	20
20	13	85	17
32	15	77	15

Note some differences between the European and American exposure techniques. The Siemens Point System takes 115 cm (42 in.) as a standard SID. The system was devised for Siemens equipment, grids, and screens. There is no American grid manufactured with 40 lines per cm (=100 lines per inch). The most useful application is the kV/mAs relationship, which is ideal for teaching and research. For example, one can start at almost any kV, go up so many points, move mAs the equivalent number of points, and have matched density exposures. It would be well for all American students to be at least aware of the Siemens Point System and how to use it.

AUTOMATIC EXPOSURE CONTROLS (PHOTOTIMERS) *

Faced with all of the possible solutions of time, kilovoltage and milliamperage, on some generators one can choose one of the 19,000 possible combinations of the three factors. These numbers in themselves reflect an extremely large potential source of exposure errors. Couple this with variations in anatomical size and density and all the other interrelated factors related to patients, and it

---

* Information in this and the succeeding section of this chapter is abstracted from T. T. Thompson, *A Practical Approach to Modern X-ray Equipment,* Little, Brown & Co., Boston, Mass. 1978.

is not difficult to see why repeat rates vary anywhere from 3 or 5 percent all the way to 35 percent.

It has long been the hope of radiology to have a device that would allow one to obtain a properly exposed radiograph regardless of patient size and condition. It has also been hoped that exposures could be reproducible from one time to another and independent of the manual selection of technique factors. Such devices have been available for a number of years, but they were better known for their variability than for their reproducibility. A device that automatically determines the amount of radiation required to produce an acceptable level of film blackness is called a *phototimer,* or *automatic exposure control.* Phototimers have their limitations, and the equipment can be used much more effectively if one recognizes these limitations.

TYPES OF PHOTOTIMERS

There are two basic types of automatic exposure control mechanisms used in conjunction with modern x-ray generators: the *photomultiplier (PM) phototimer* and the *ionization chamber.* The photomultiplier type measures light intensity and the ionization chamber measures an ionization current. Phototimers may be located in front of the image receptor (entrance types) or behind the image receptor (exit types). In most cases, the image receptor is the film cassette.

Photomultiplier phototimers

The detector of most photomultiplier phototimers is made from thin, segmented sheets of Lucite that have been covered with black paper, except in the region that will be used to detect radiation (Figure 8-3). One of the properties of Lucite is that it will transmit light in a manner similar to that of a fiberoptic bundle. The sheets of Lucite, called *Lucite paddles,* are coated with a phosphor which, when struck with radiation, will produce light in proportion to the intensity of the radiation. This light is transmitted by the Lucite to an output area called a *light gate* and coupled to a photomultiplier tube. The photomultiplier converts the light photons into electrical signals, which can be amplified and used to control the termination of the exposure. When the automatic exposure control senses the amount of radiation required for proper film density, the exposure is terminated. Modern silicon-controlled rectifier circuits allow maximum reproducibility at very short exposure times.

In an entrance-type PM (photomultiplier) phototimer, that is, one which is located in front of the cassette, three apertures are normally available. The size of the aperture varies according to the manufacturer, but the average size is 100 cm². Each aperture is connected via the Lucite sheet to a light gate. There may be one PM tube per gate, or there may be one tube for all three gates. Electrical or mechanical switches are provided on the generator console to allow the opera-

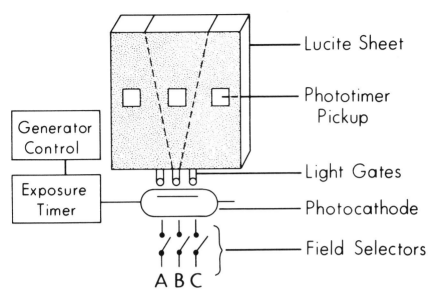

Figure 8-3. A modern photomultiplier-type automatic exposure control. As radiation strikes the phototimer pickup, light photons are produced in the phosphor coating the pickup paddle. This light is transmitted throughout the Lucite sheet and to the light gates. At the light gate the photocathode detects this light and generates an electrical signal, which is fed into the exposure timer. The exposure timer is adjusted so that when the proper amount of current is received from the photocathode, it indicates to the generator that the exposure is to be terminated.

tor to choose any combination of fields or apertures, up to a total of seven combinations.

One of the major problems with PM phototimers has to do with the response of the phototimer to changes in kilovoltage. At low kV ranges (50–60 kV) there is proportionately more x-ray energy absorbed by the pickup paddle than at high kV levels, leaving less energy to expose the x-ray film; the result is that the exposure is terminated before the proper film density is obtained. Thus, more x-ray energy is required at low-kV levels to produce an acceptable film density than at higher kV levels. In technical jargon, this problem is that of "being oversensitive at low kilovoltage." The problem is overcome by using what is known as *kilovoltage compensation,* a method in which the kilovoltage being used for a specific exposure is sensed and an electrical signal is sent to the detector to raise or lower its sensitivity, thereby compensating for variability of absorption of the x-ray beam by the detector. At low kV levels the phototimer detector shows less sensitivity, allowing more radiation to reach the film before the exposure is terminated.

Photomultiplier phototimers are the most commonly used exit-type detectors. In many cases a single aperture is used, the size of the aperature varying with different manufacturers. The aperture is generally larger in exit-type than in entrance-type phototimers. The main advantage of a single-field phototimer is that it is generally much less expensive than a multiple-field type and is also less trouble to maintain. Another advantage is that thicker phosphors can be used on the pickup paddles, allowing a stronger signal to be generated. In addition, the same phosphor that is used in the film/screen combination can be readily used in the pickup paddle without the operator having to worry about absorption and scattering of the x-ray beam. The disadvantage of the single-field detector is not electromechanical, but is primarily technical in nature: *its use requires far greater attention to be paid to the proper positioning of the patient over the detector sensor.* Exit-type detectors have been de-emphasized by most manufacturers in recent years, primarily because of the need for special control of cassette opacity.

Ionization chamber phototimers

An ionization chamber (Figure 8-4) is composed of extremely thin sheets of foil, usually aluminum or lead, in which are enclosed radiation-sensitive ionization chambers. The ionization chamber contains a gas, usually air under normal atmospheric pressure, that ionizes in direct proportion to the amount of radiation striking it. When the air is ionized, it forms a complete electrical circuit, and the current

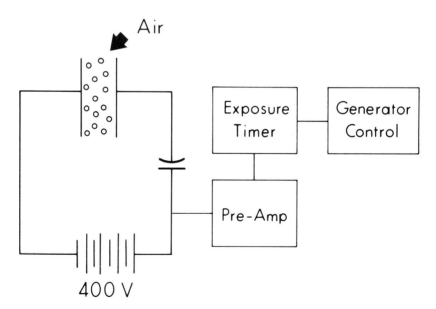

Figure 8-4. Ionization-type automatic exposure control. As radiation strikes the gas contained in the exposure control, ionization of the gas occurs, which produces a minute electrical current used to charge a capacitor. When the capacitor is charged to a predetermined level, it is discharged, and the signal produced by the discharge controls the exposure, as with the PM-type automatic exposure control. Ionization chambers are almost universally located in front of the image receptor.

from this circuit can be amplified and used as a signal to terminate the exposure. Except in mammography, ionization chambers are almost universally used as entrance-type phototimers. Automatic exposure controls based on ionization chambers are not phototimers in the strictest sense, but the term "phototimer" is almost universal today for both classes of detector.

LIMITATIONS OF PHOTOTIMERS

When phototimers are used it is necessary that the films, screens, and cassettes be standardized for speed and absorption. Inconsistency in the absorption characteristics of one screen or cassette as compared with another cannot be tolerated for any type of automatic exposure control, but particularly not for exit types. If either the film speed, screen speed, or absorption of the cassette varies from the standard set for a particular piece of equipment, the phototimer will not provide a correct film density. Phototimer cassettes are not absolutely essential for entrance-type phototimers, but films and screens must be standardized if one wants to avoid or eliminate operator error. Thus one of the largest problems with phototimers is not with the phototimer itself, but with the variability of supplies used in conjunction with the device.

Another problem is related to the size of the automatic exposure device. When an exposure detector is placed between the tabletop and the Bucky tray, it takes up space. This additional need for space increases the object/image receptor distance, thus increasing magnification distortion. The insertion of the exposure control device into the incident x-ray beam also produces scatter radiation, which decreases radiographic contrast. However, the loss by way of unsharpness and diminished contrast is more than offset by the reproducibility offered by the exposure device and the automation afforded by the use of such devices.

PERFORMANCE OF AUTOMATIC EXPOSURE CONTROLS

The Bucky automatic exposure control is not an answer to all the exposure problems that face a machine operator. However, if the limitations of the equipment are understood, it is easier to obtain satisfactory results with almost any automatic exposure control mechanism than by continually facing the frustrations that commonly occur with this equipment. One should be aware that no automatic exposure termination device is perfect, even with the solid-state electronics now available to the industry. The automatic exposure control circuits must meet the regulations of the Department of Health, Education, and Welfare, which state that the generator must automatically terminate an exposure at 600 mAs for 50 kV or greater, or at 2000 mAs for kV under 50. The performance standards also state that an exposure must be reproducible within 10 percent from exposure to exposure.

As far as exposure linearity variation is concerned, HEW restricts variation in the average ratio of exposure to the indicated milliamperage product (millirads/mAs) obtained at any two consecutive tube current settings to no more than 0.10 times their sum. This means that if 6 millirads/mAs is required for an exposure at 100 mA at 100 ms and another exposure is made at 200 mA at 50 ms, the second exposure must be within 4.8–7.2 millirads/mAs. For practical use, the mAs cannot vary more than ±10 percent.

Minimum response time

The automatic exposure timer is a component of a system the purpose of which is to provide an electrical signal to terminate an exposure. The time required for an automatic exposure control device to terminate an exposure after a proper amount of radiation has been sensed is called the *minimum reaction time*. There are a number of components that constitute this reaction time. The first of these is the reaction time of the automatic exposure control mechanism itself, which may be in the range of 2 to 3 ms. In addition, some time is required to operate mechanical relays. In older equipment this time can be stated to be about 8 ms. The third factor has to do with the electrical connections between the automatic exposure control and the generator. Without forced extinction, i.e., forced termination of the exposure, an additional variable of 8 ms must be added for the voltage waveform to cross zero potential. Repeatability of these older automatic exposure controls is limited to 30 ms. With forced extinction and proper anticipation circuitry, the total system reaction time can be within 1 or 2 ms. Automatic exposure controls used with some generators can provide density tracking down to 1 ms. However, 2 ms is a far more commonly used specification for those units.

Checking the minimum reaction time of an automatic exposure control device usually requires the aid of an oscilloscope. If one calculates what nominal exposure factors would be required for a routine radiographic exposure, taking into consideration screen speed, milliamperage station, and kilovoltage, an estimate of the exposure time can be made. If the exposure time is less than 20 ms, it is wise either to lower the kilovoltage or the milliamperage station or to use lower-speed screens to make certain that the exposure time will be longer than 20 ms (unless adequate controls, such as forced electronic termination, are available for shorter repeatable exposures).

Repeatability of short-time exposures

An automatic exposure control should be capable of making repeated exposures at short times with no more than ±10 percent variation in film density for an average of five exposures. To test this repeatability, one should place an aluminum step wedge over the detector aperture and manipulate the exposure factors (kV or mA) until a short exposure time is reached (about 50 ms will do). Five exposures * should be made using these technique factors; the density between the films should not vary more than ±10 percent. For practical purposes the film density should not vary more than one density step on the step wedge.

Repeatability of long-time exposures

An automatic exposure control should meet the same requirements for long-time exposures as it does for short-time exposures. Long-time repeatability is tested in the same manner, except that technique factors are chosen so that the response time will be about 1 sec. Exposures longer than 1 sec are seldom used in radiography except for tomography.

Response to changes in tissue thickness

An automatic exposure control in good working condition should be able to compensate for varying thicknesses of the anatomical parts being radiographed. To check this response, one should place an aluminum step wedge over the detector aperture and make an exposure at 70 kV. One should then repeat the procedure, adding a piece of old fluoroscopic apron, aluminum sheeting, or Lucite blocks between the x-ray beam and the automatic exposure control detector. The procedure should be repeated several times until the equivalent of five layers of apron are filtering the x-ray beam. Ideally, water should be used when the exposures are made; a large container should be placed with an inch or so of water in it, and water should be added gradually between exposures to simulate patient tissue. This procedure should be repeated for the 90-kV and 120-kV levels. Film density at all kV levels and for varying "patient thicknesses" should not exceed 10–15 percent over the entire range of thicknesses. For ideal comparison a step wedge can be placed outside the measuring field. Figure 8-5 illustrates an acceptable method for accurately checking phototiming tracking.

TECHNICAL CONSIDERATIONS

Assuming that an automatic exposure control meets the performance criteria noted above, there are additional factors that must be considered (some of which have already been alluded to). One cannot overemphasize the importance of having matched screens, films, and cassettes for use with any type of automatic exposure control device. If the automatic exposure control is presented with variables other than those for which it has been calibrated, the expected film density will not be obtained. Electronic devices can be made to track evenly, but with the non-linear responses of screens, it becomes more difficult to track evenly. The processing of the radiographic film is just as crucial a factor for

---

* WARNING: Go slow with these multiple exposures; keep anode thermal capacity in mind.

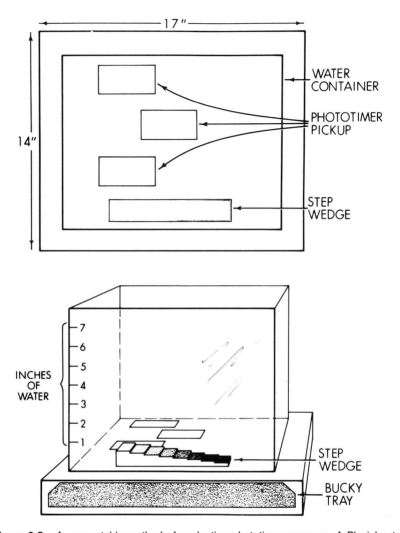

Figure 8-5. An acceptable method of evaluating phototimer response. A Plexiglas tank is placed over the phototimer pickup. A standard step wedge is placed in the tank to be adjacent to the pickup stations. Water, which serves as an absorber and scatterer, is added to the tank in increments. For every inch of water added to the tank, an exposure is made. The film density from one exposure to the next should be within one step-wedge step of the other exposures.

films exposed by the automatic exposure control mechanism as it is anywhere else in the radiology department. Sensitometric control of the processing must occur before any adjustments are made in the x-ray equipment, particularly with the automatic exposure control.

The density control buttons or rotary knobs provided with the automatic exposure control can be used to increase or decrease the sensitivity of the detector. These controls normally allow a range of density controls from −50 percent, −25 percent, 0, +25 percent, to +50 percent. On some equipment these controls are labeled −2, −1, 0, +1, +2, but the effect is about the same. If a film is light or dark at different technique levels as demonstrated by the kilovoltage linearity test described above, p. 209, the operator may be able to use the density controls to bring the film density into a useful range while waiting for a serviceman to adjust the automatic exposure control properly.

The use of automatic exposure controls by student radiographers has long been the bane of training instructors, since the students do not have to select appropriate technique factors. While the use of automatic exposure controls may discourage students from learning correct exposure techniques, it is the experience of most manufacturers and users of automatic exposure controls that the positioning of the patient over the measuring field aperture is far more critical than when conventional techniques are used. The automatic exposure device cannot replace a well-trained technologist.

Selection of the aperture field is as important as correct positioning of the anatomical part over the aperture. For example, with a three-field pickup, choosing the middle aperture will allow a proper exposure of the mediastinum to be made on a chest radiograph, but the lung fields will be overexposed. Choosing the right-field aperture may or may not allow a correct exposure of the lungs, depending on whether or not the cardiac shadow will interfere with the sensing of the correct film density. The selection of the left-field aperture will allow a correct exposure of the lung fields, provided there is no gross pathology in the lung. Unless one is using very high kV techniques (125–150 kV), a chest radiograph performed with an automatic exposure control will not correctly expose both mediastinum and lung parenchyma. The same holds true for the lateral chest radiograph. Positioning the lateral thoracic spine over the aperture will allow a correct exposure to be made of the thoracic spine, but will overexpose the lung fields; selection of an aperture for the lung fields will make likely the underexposure of the upper thoracic spine. Placement of the anatomical part over the aperture in such a way that only a portion of the anatomical part is sensed will result in underexposure or overexposure, depending on how much of the anatomical part is sensed by the detector.

Collimation is just as important with automatic exposure control devices as it is with standard exposures. Scattered radiation resulting from widely collimated areas will reach the detector and terminate the exposure before a proper film density is detected.

Although the capabilities of automatic exposure controls will vary widely from manufacturer to manufacturer, if the operator knows the capability of the equipment, it is possible to make and reproduce acceptable exposures. In the meantime, manufacturers are continuing to perfect a better and more reproducible automatic exposure control mechanism.

ANATOMICAL PROGRAMMERS

A basic problem in the selection of proper exposure factors is the extremely large number of possible combinations available. Some thirty years ago Picker

Cahoon's Formulating X-Ray Techniques

Medical Systems of Cleveland, Ohio, introduced a method whereby the operator chose only one technique factor, and the generator was preprogramed to allow a present kV-mA-time for that technique factor. This first anatomical programer never really caught on, primarily because no good automatic exposure device was available then. In recent years European manufacturers have offered anatomical programing as a feature of their generators used in conjunction with automatic exposure controls. This method is widely used in Europe as well as in many institutions in the United States.

Like its predecessor, the new anatomically programed module is an accessory for the generator. Equipment may be purchased that has anywhere from 8 to 40 or more anatomical programs (Figure 8-6). The purchaser decides within a given parameter what technique factors are to be dedicated to a particular tech-

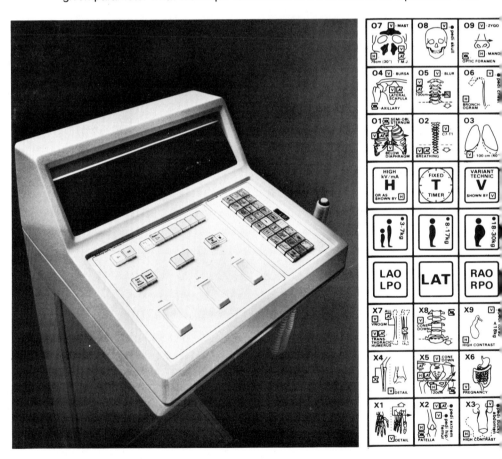

Figure 8-6. Modern x-ray generator console with technique management system. Each button located on the program is pre-set for kilovoltage, milliamperage, time, and focal spot. To obtain a correct exposure, the technologist simply positions the patient properly over the appropriate phototimer pickup(s), and activates the anatomical programmer and exposure control. The use of microprocessors can greatly extend the flexibility of a system such as this. (Courtesy, General Electric Co., Milwaukee, Wisc.)

nique button. For example, the technique factors for a radiograph of an elbow might be 55 kV, 100 mA, 100 ms, small focal spot. When the technique button for "elbow" is engaged, the generator is automatically set for those factors. The anatomically programed generator is best used in conjunction with automatic exposure devices.

Programing of complete special-procedure examinations is already available with some complex systems and will most likely be utilized more in the future. An ideal application of anatomical programing is in conjunction with gastrointestinal spot-film radiography. With conventional equipment, either the radiologist or the technologist has to change the generator control when going from spot-filming of the gallbladder to spot-filming of the gastrointestinal tract. With anatomical programing provided at the spot-film device, the radiologist would simply have to push the "gallbladder" technique button; when filming of the gallbladder was completed, the "GI" button would be activated. Normally, lower kilovoltage (70 kV) is used when radiographing the gallbladder as compared with the high kilovoltage (120–150 kV) required to penetrate the barium used in gastrointestinal fluoroscopy.

Microprocessors—miniature computers—will soon begin to replace the relays and discrete transistors in the logic circuits of anatomical programers. The microprocessor will provide access to a larger but inexpensive memory on a single silicon chip. From stored data on this chip, the optimal milliamperage, safety time, and even non-automatic exposure control fixed times can be added to the list of factors that are automated. Programed selection of the tube load (kV and mA) may eliminate the need for falling load circuitry.

SUGGESTED READING

"Exposure Initiation/Termination Mechanisms and Automatic Exposure Timers in Diagnostic Radiology." (Proceedings of the Society of Photo-Optical Instrumentation Engineers. *Application of Optical Instrumentation in Medicine,* IV, 70(1975):165.
"Automatic Exposure Controls," in T. T. Thompson, *A Practical Approach to Modern X-ray Equipment* (Little, Brown & Co., Boston, 1977).
Eastman, Terry R. "Automated Exposures in Contemporary Radiography." *Radiologic Technology* 43(1971):80–83.

REVIEW QUESTIONS

1. In the Du Pont Bit System for Technique Conversions, subtracting 1 bit results in

a. 100% change in density   c. 50% change in density
b. 75% change in density   d. 25% change in density

Cahoon's Formulating X-Ray Techniques

2. Using Tables 8-1 and 8-2, determine the following technique conversions:

Old kV	New kV	Old mAs	New mAs
64	80	30	?
88	120	5	?
70	?	45	10

3. In Figure 8-1 (p. 193), what would the new mAs be for a chest exposure originally made at 65 kV, 20 mAs?

4. Measurement of light intensity is the basis of what type of automatic exposure device?

    a. Photomultiplier type
    b. Ionization chamber
    c. Bucky chamber
    d. Exit type
    e. Entrance type

5. In reference to an automatic exposure control, minimal response time is defined as

    a. The time required for a film to respond to light
    b. The response of screens to changes in kilovoltage
    c. The minimal time for an exposure to be terminated after a proper amount of radiation has been sensed
    d. The reaction time of the phototimer itself
    e. The relay closure time

6. A microprocessor is a

    a. Table top film processor
    b. Automatic film processor for minifilms
    c. Computer using microsize logic circuits
    d. Leitz microscope
    e. Minute nerve controlled by the autonomic system

7. An anatomical programmer is

    a. An accessory for programing exposure factors
    b. A program for learning anatomy
    c. A self-instructional booklet for learning electronics
    d. The person who develops technical charts
    e. A type of phototimer

8. When using the Siemens Point System, the addition of 3 points would

    a. Quarter mAs
    b. Halve mAs
    c. Double mAs
    d. Triple mAs

FORMULATING TECHNIQUE CHARTS

There are two basic systems of radiographic exposure technique in common use today. One of these—the fixed-kilovoltage technique—was developed by the late Arthur W. Fuchs and employs a fixed or optimum kilovoltage for a given anatomical part. The selection of the kilovoltage used in the examination is based on the penetrating quality of the x-ray beam for the average adult patient. The mAs is used as a variable, all other factors remaining constant. The variable-kilovoltage exposure system, like most systems, requires that the SID (source/image-receptor distance, focal-film distance) remain constant, but the kV and mAs are varied to compensate for variations in thickness and density of the part being radiographed.

The fixed-kilovoltage method is simple to use and maintains a constant level of contrast for a particular examination. Its simplicity of operation and consistency in radiographic quality promote greater technical accuracy. Since the kilovoltage used in the examination is adequate to penetrate all anatomical parts in the field, more detail is visible.

One also has a moral obligation to the patient when deciding which exposure system to use. The exposure of the patient to radiation is likely to be greater with a variable-kilovoltage technique than with a fixed- or high-kilovoltage technique.

It is not possible to construct a technique chart that is precisely workable on all types and makes of apparatus, because of varying degrees of calibration efficiency. The best that can be accomplished is a compromise in which all factors but one are reduced to constants. The variable factor employed should have only one function—to influence radiographic density. No technique chart is a cure-all or the answer to good radiography. It is only a scientific aid in that direction. Training in positioning, the application of suitable immobilization devices, and cones and grids (when employed), all contribute to the quality of the radiograph. Clinical experience in estimating the relative x-ray absorption characteristics of the patient as altered by disease, trauma, or age also has a decided bearing upon the results. Adjustments for relative absorption can be approximated by using mAs values in the next lower or higher thickness category.

The exposure charts in this chapter, which are designed for use with single-phase x-ray equipment, list suggested exposures for various routine projections. It may be found practical to adapt these factors to the routine radiographic work required by your department.

PREPARATION FOR FORMULATING THE TECHNIQUE CHART

Before commencing with the development of the technique chart, one must make sure that conditions are standardized, so that when the chart is completed, it will be applicable to the department in which it is to be used. Standardizations of conditions are summarized below.

Radiographic equipment. Present-day radiography requires standardization of apparatus and technique in order that all operations performed may entail a minimum of time and yield results of the best quality. The equipment should be calibrated to assure that there is a proper and consistent output in radiation at all ranges of kV and mAs.

Grids. Ascertain what grids are available in the department. Note that not all grids with the same grid ratio and lines per inch necessarily absorb or transmit the same amount of radiation. Particularly, check the grids in Buckys to make sure that at least the grid ratio and lines per inch are standardized in the department. In addition, the grid fronts may be made from different materials which have different absorption characteristics.

Screen speed. Check to be sure that the screens are standardized in the department. A medium-speed screen manufactured by one company may not necessarily be equal in speed to a medium-speed screen manufactured by another company. Also make sure screens are clean and undamaged.

Collimator light-field/radiation-field alignment. Make certain that the light field of the collimator is aligned with the field of the x-ray beam, not only at the standard 100-cm distance, but particularly at the 183-cm distance.

Elimination of variables. When formulating the x-ray technique chart, make certain that the same cassette, screen, and grid are used for each exposure so that variables can be partially eliminated. In addition, exposed films should be processed through the same processor, in which processing has been optimized.

FIXED-KILOVOLTAGE TECHNIQUE

In this technique the kilovoltage is optimized for the part being radiographed, and the mAs is varied to compensate for patient thickness. While mAs controls density, the kV is fixed at a level which will penetrate the part irrespective of its size. The fixed-kV system of technique has become popular since the kV controls or affects many factors—contrast, density, quality of radiation, production of secondary radiation, scattered radiation, and exposure latitudes, to name six of them. All six factors may vary with the variable-kV technique, but except for density, remain constant with the fixed-kV technique. Thus it becomes difficult to maintain a balanced scale of contrast, for when voltages are changed to compensate for part thickness, the contrast scale also changes.

The beginning student in radiography has difficulty in understanding or describing the quality of the radiographs he produces, whether good or bad. For that reason, early in his training he must become acquainted with the kind of radiograph that is most useful to the radiologist—one that possesses an acceptable diagnostic quality. The student cannot acquire that knowledge of acceptable film quality except by actually making such radiographs under careful guidance. Familiarity with satisfactory radiographs, based on experience in making them, instills a self-confidence that is of immeasurable value in the field of radiography. The fixed-kV technique, with its more consistent results, allows the student to learn earlier. Wherever practical, the exposure factors in the fixed-kV technique are reduced to constants, thereby eliminating many possible sources of error. This system enables the student to quickly become acquainted with radiographs of better than average quality that are produced in the course of instruction. It precludes floundering about, seeking exposure factors for making a particular radiograph the quality of which one is often incapable of judging in the early stages of training. The fact that the fixed-kV system requires the elimination of nonessentials or variables from the technical procedure accounts for its success, since it provides a clear-cut technique, assures the technologist of better than average results, and is easily understood and applied by the beginner as well as the experienced radiologic technologist.

With some experience, the technologist learns to estimate the thickness and is able to classify the part as *small, average,* or *large.* Since the average classification predominates, about 90 percent of all patients may be placed in this category. The basic constants of kV and mAs are so few in number that they are easily memorized; and for ready reference, the factors may be written on a small card. Although the fixed technique is simpler to use than the variable technique, the technologist should still measure part thickness, since determination of part thickness is still a common problem in the repeat examinations.

The employment of the fixed-kV technique will consistently produce radiographs of good quality with a minimum of effort. Since the method is standardized, shifting of personnel from one room to another should not interfere with the uniformity of results; each technologist trained in it should produce a radiograph of the same quality and type as his predecessor. The radiographic densities produced by this method are uniform—a point of considerable diagnostic value, because any differences from the normal density may be attributed to pathological changes within the tissues. Duplication of results is easy to obtain in follow-up.

For a given thickness of a particular body part, the wavelengths of the x-rays employed to penetrate the tissues must be adequate. Since the kilovoltage governs the wavelength and penetration, all that is necessary for securing a satisfactory image is to have an optimum number of x-rays reach the film. The milliamperage and the time control the number of x-rays—the intensity of radiation—and proper adjustment of these factors will produce adequate exposure of the radiographic film. When the x-ray wavelength is correct, less mAs is usually needed for the exposure that is habitually employed for the same purpose with lower kilovoltages. This is most important where conservation of tube life is desirable, for it is the mAs that commonly destroys tubes—not kV.

In determining the kV to be used in a fixed-kV system of technique a premise must be established as to the diagnostic acceptability of the image. The scale of contrast should be such that all anatomical details are readily visible. This, of course, depends on the penetration. Each part should be adequately penetrated. Only a minimum of secondary radiation fog should be tolerated. The average density level should be such that the majority of densities are translucent when the standard type of illuminator is used.

HOW THE FIXED KILOVOLTAGES WERE DETERMINED

The determination of fixed kV for various projections involved experimentation and testing. Most of the experimentation was done at a time when there was little concern for radiation safety and radiation exposure to the patient. The process described below was developed by Fuchs.

1. A series of exposures were made of a particular part, employing a given projection. The radiographs were made at various kV, usually 40 to 100, in 10-kV steps. At first, a number of values were employed, but later the kV's were narrowed to three or four for trial. The mAs value used with each kV was adjusted so that the average overall densities of the radiographs were approximately the same. The densities for the same anatomical area were balanced for each radiograph.

When it was determined that a given kV would be adequate, it was tried on a large number of patients, using a single basic mAs value. In some instances the radiographic densities were adjusted to suit the average range of tissue densities. As soon as the radiographs assumed a measure of uniformity, the kV and basic mAs factors were established as constants.

2. Measurements of tissue for a given projection were carefully made. Calipers were employed and skin-to-skin measurements were made without compression of tissues. When these measurements were obtained, correlation with anthropometric data was secured so that frequency and reliability of the measurements could be assured.

The frequency of measurement to establish the average thickness ranges was established by Fuchs in the late 1930s and 1940s. However, the American population has since grown in height and breadth. Body-part sizes have increased since Fuchs's data were published. This requires an increase in the kV for optimum penetration. The kV established by Fuchs was selected to suit the nature of radiography to be done—whether screen, direct exposure, or screen-grid exposures—and the greater frequencies were then tested radiographically, using the kV selected as optimum for the projection.

When measurements were encountered that were greater or less than average, some adjustment in the mAs was necessary. However, the kV was to remain constant. Borderline thickness ranges were established. In some instances, compensation could be made by halving the mAs for parts measuring less than the average and doubling the mAs for parts greater than the average. This could usually be done where range in part thickness was low, as with the posteroanterior view of the hand.

As the range in part thicknesses became higher, refinements in the thickness divisions had to be made. Taking the basic mAs as x, then the halfway point between x and $2x$ would be $1.5x$. Also, with x and $0.5x$, the halfway point would be $0.75x$. For example, if 80 kV was initially established for the chest projection (PA) with a basic mAs of 3.3 at 183 cm (72 in.) using screens, then these factors were used on 82 percent of all adult patients entering the department and measuring 20 to 25 cm. Cases measuring 25 to 29 cm were exposed with $2x$ mAs; those measuring 16 to 20 cm were exposed with $0.5x$ mAs. These figures were later revised so that for a 21 to 24-cm range, x mAs was used; $1.5x$ mAs was used for 24 to 27 cm; and $2x$ was used for 27 to 30 cm. For the 18 to 21-cm group, $0.75x$ was used, and $0.5x$ was used for the 16 to 18-cm group. The same procedure was employed for all projections.

Adjustment of the kilovoltage should always be made when a new projection is established. Just as soon as new conditions are introduced into a standard projection, it should be considered a new one, and factors should be laid down to satisfy the requirements of that projection.

THE FIXED-KILOVOLTAGE TECHNIQUE CHART

The fixed-kV exposure guide is divided into five main areas: chest, axial skeleton, cranium, abdomen, and peripheral skeleton. Table 9-1 illustrates a typical technique chart which lists the part to be examined, projection, thickness of part to be radiographed, kV level, and suggested mAs. The chart also indicates whether or not a grid is used, and gives the SID and focal spot. Changing the kV of the chart is not recommended, in keeping with the basic philosophy of fixed-kV techniques. The recommended grid is an 8 : 1, 100–103-line, aluminum interspaced. Standardized automatic processing is recommended.

Selecting the milliampere-second

Referring to Table 9-1, the average lateral skull will be from 15 to 18 cm in width and the optimum kV to penetrate the lateral skull is 80 kV. To determine what mAs is required for the examination, make a series of four radiographs of a skull phantom placed in the lateral position. For this determination, the first

Cahoon's Formulating X-Ray Techniques

Table 9-1. Fixed-kilovoltage technique guide

Grid 10:1/100–103 line
aluminum interspaced
Medium-speed screens [a] and films

Part	Projection	cm Range	kV	Sug- gested mAs	Selected mA	Selected Time	Grid	Focal spot	SID (cm)	Film size, in.
Skull	Lateral	15–18	80	20			Yes	S	100	10 × 12
	Anterior(PA)	18–21	80	50			Yes	S	100	10 × 12
	Half-axial (modified Towne)	18–21	80	50			Yes	S	100	10 × 12
	Basilar (submento-vertex)	23–26	80	100			Yes	S	100	10 × 12
	23° Caldwell	18–21	80	50			Yes	S	100	10 × 12
	Stenvers	16–19	80	40			Yes	S	100	
			80	20			Yes	S	90	8 × 10
	Mastoid lat. (Law)	13–17	80	20			Yes	S	75	
			80	30					100	8 × 10
	Optic foramina	18	80	80			Yes	S	75	
				100			Yes	S	100	8 × 10
Sinuses	Waters (PA maxillary)	18–22	80	50			Yes	S	90	
			80	50			Yes	S	100	8 × 10
	Lateral	11–14	80	12			Yes	S		8 × 10
	Open-mouth sphenoid	21–24	80	50			Yes	S	90	
			80	60			Yes	S	100	8 × 10
	PA(frontal)	18–22	80	50			Yes	S	90	
			80	60			Yes	S	100	8 × 10
Mandible	Axiolateral	9–13	70	5			No	S	90	
			70	6.6			No	S	100	8 × 10
	PA	12–15	80	25			Yes	S	100	8 × 10
Nose	Lateral	—	65	5			No	S	100	
Chest	PA	<14		1.66			Yes	L	183	14 × 17
		15–18		3.3			Yes			14 × 17
		19–22	80	6.66			Yes			14 × 17
		23–26		10			Yes			14 × 17
		27–30		13.3			Yes			14 × 17
		31–34		16.6			Yes			14 × 17
	Lateral	<26	90	7.5			Yes			14 × 17
		27–31		10			Yes			14 × 17
		32–36		15			Yes			14 × 17
	Oblique	19–22	86	5			Yes			14 × 17
		23–26		10			Yes			14 × 17
		27–30		15			Yes			14 × 17
Ribs	Above diaphragm	21–24	76	15			Yes	L	100	Select for coverage

Table 9-1. cont.

Part	Projection	cm Range	kV	Sug-gested mAs	Selected mA	Selected Time	Grid	Focal spot	SID (cm)	Film size, in.
	Below diaphragm	17–21	76	80			Yes	L	100	Select for cov-erage
Cervical spine	Lateral and oblique	11–14	80	10			No	S	183	8 × 10
	AP	11–15	80	12.5			Yes	S	100	
Thoracic spine	AP	20	80	60			Yes	L	100	7 × 17
	Lateral	30	80	100			Yes	L	100	7 × 17
Lumbar spine	AP	20	80	80			Yes	L	100	10 × 12
	Lateral	30	100	150			Yes	L	100	10 × 12
	Oblique	24	80	200			Yes	L	100	10 × 12
Pelvis	AP	20	80	80			Yes	L	100	14 × 17
Fingers	AP	1.5–3	60	20			Non-grid	S	100	8 × 10
Hand Wrist	AP	2–5	60	2.5			detail screens	S	100	8 × 10
Elbow	AP	7–8	70	2.5				S	100	8 × 10
Shoulder	AP	13–14	76	20				S	100	8 × 10
Foot	AP	6–7	60	5				S	100	8 × 10
Ankle	AP	9	70	10				S	100	8 × 10
Knee	AP	11–12	70	30				S	100	8 × 10

[a]Although this technique chart shows the use of medium-speed calcium tungstate screens, it is recommended that high-speed screens be used; technique conversion is made by halving mAs.

exposure should be a 200 mA, small focal spot, and exposure times of 50, 100, and 200 ms, which will allow 10, 20, and 40 mAs respectively. (The 100-mA station could also be used, but in that case the exposure time would be doubled.) In any case, the first exposure should be approximately the same as the recommended kV and mAs, with the test exposure subsequently halved and doubled. Process the radiographs and select the one that meets the standard of quality expected. Suppose that radiograph no. 3 is chosen as the optimal film. In this case, where the exposure was made at 80 kV, 200 mA, small focal spot, 200 ms, the radiograph of any non-pathological skull which measures 15, 16, 17, or 18 cm in breadth will require these technique factors. For any lateral skull measuring less than 15 cm, cut the mAs in half; for skulls measuring more than 18 cm, double the mAs.

Once these technique factors have been formulated, the entire section on the chart can be filled in by developing a ratio between the suggested mAs and the actual mAs found by the above-mentioned method. For example, if the technique chart has a suggested mAs of 25 for a lateral skull, and by experimentation you find that 40 mAs is required for your particular equipment, then the ratio is $^{40}\!/_{25} = 1.6$. Therefore, the suggested mAs will have to be increased 1.6 times for all views. The same process can be used to fill in the actual or selected mAs for all other views listed in the chart. As actual exposures of patients are made, the mAs selection can be refined for all views and projections and for each radiographic room.

Technique conversions for fixed-kilovoltage techniques

Table 9-1 was developed against fixed standards. Sometimes it is necessary to use other methods than these. The following technique conversions will be helpful:

1. From medium-speed calcium tungstate screens to high-speed calcium tungstate screens: cut mAs in half.
2. From medium-speed calcium tungstate screens to detail calcium tungstate screens: double the mAs. (Some detail screens require more or less exposure depending on brand.)
3. To go from a 10 : 1/100–103-line grid to an 8 : 1/80-line grid: no technique change is required. For an 8 : 1/100–103-line grid: decrease mAs by 25 percent.
4. To go from non-grid technique to a 12 : 1/80-line grid: add 20 kV.
5. To convert from single-phase to a 3-phase, 12-pulse, or constant potential generator: cut the mAs in half. To convert from a single-phase to a 3-phase, 6-pulse generator: cut the mAs by 40 percent.

VARIABLE-KILOVOLTAGE TECHNIQUE

Because of the greater exposure of the patient to radiation, the question can be asked, Why are variable low-voltage techniques used today? The reason is that variable-kV techniques allow greater contrast than fixed-kV techniques. The function of contrast is to enhance the visibility of detail. Accordingly, many radiologists with a preference for short-scale contrast limit their technologists to techniques of the 40-kV and 90-kV range. In the 1920s and 1930s, low-voltage techniques were used because of equipment limitations and because devices such as 8-to-1 and 12-to-1 grids which provide adequate control of secondary and scattered radiation fog were not available. Radiologists became accustomed to interpreting films that were of high contrast. However, most low-voltage films have areas which are not completely penetrated and which thus lack silver deposits. Basically, areas lacking silver deposits are diagnostically useless; yet because of the pictorial quality of low-voltage films, many subjectively feel that they are superior in radiographic quality. It is surprising to find that even today lateral lumbar spines are radiographed at low (75) kV rather than high (100) kV.

A technologist can do remarkably good work with the variable-kV technique, although the radiation exposure is higher than with a fixed-kV or high-kV technique, and the contrast scale changes with each part thickness. No matter what radiographic exposure system he chooses, the radiologic technologist must be certain of the calibration of the radiographic equipment, since it becomes impossible to formulate standardized techniques without proper calibration of equipment.

FUNDAMENTAL CONCEPTS OF THE VARIABLE KILOVOLTAGE TECHNIQUE

With the variable-kV technique, the mAs remains constant and the kV changes according to the centimeter thickness of the patient. Normally one would expect the technique to increase in increments of 2 kV per centimeter thickness. The basic formula to be used in determining the appropriate kV is 2 × the centimeter thickness plus 30 kV. Thus a body part which measures 15 cm would require $(2 \times 15) + 30 = 60$ kV, and for thickness greater or less than this 15 cm, we would require an increase or decrease of 2 kV for each centimeter change. The mAs is so chosen as to provide an optimum film density.

The above concept presupposes that all equal tissue thicknesses will have equal x-ray absorption—which we know is not true. For example, an anteroposterior (AP) shoulder and a lateral skull have similar thicknesses, but the skull is much denser than the shoulder and will require comparably more kV for adequate penetration. For this reason, i.e., variability of tissue absorption for the same centimeter thicknesses, anatomical parts have been arbitrarily divided into groups which not only have similar thicknesses, but also similar densities. Basically the groups are these:

1. Extremities
2. Skull
3. Sinuses
4. Trunk and pelvis
5. Lateral lumbar spine
6. Lateral thoracic spine
7. Chest
8. Gastrointestinal tract
9. Lateral cervical spine

Thus, although the basic formula for determining kilovoltage will work, it has to be optimized further by considering the tissue density as well as the tissue thickness.

Besides considering the variations in tissue thickness and density, some consideration must be given to disease processes. Patients are arbitrarily divided into three classes: Class A, emaciated or easy to penetrate disease processes; Class B, normal patients; Class C, disease processes or muscular patients difficult to penetrate. As a general rule the optimized mAs is decreased 30 percent for Class A and increased 30 percent for Class C. (The patient problem will be further discussed in Chapter 10.)

VARIABLE-KILOVOLTAGE TECHNIQUE CHART

Table 9-2 provides data to be used as a starting point for developing a variable-kV technique chart. To convert these data for use in a department, follow these guidelines:

1. Select a phantom of average size and start with any group of parts of similar density. The lateral skull will be used as our starting point. Measure the subject in the lateral projection. On the average, this measurement will be 15 cm. Multiply the centimeter thickness by 2 and add the figure 30 to the product. In this manner we arrive at the figure 60, our base kV. In other words, a 15-cm lateral skull would require 60 kV; a 16-cm skull would require 62 kV; a 14-cm skull would require 58 kV.

2. Select the mAs thought to be most appropriate. As a starting point choose 100 mAs.

3. Select the SID which will be appropriate for the examination. Generally this will be 100 cm, except 183 cm for chest exposures and 91 cm for most skull radiographic units.

4. Decide whether a grid should be used. As a general rule, exposures of body parts measuring more than 12 cm should be made with a grid.

5. The technical factors we will use, then, are these:

SID:	100 cm
Grid:	10 : 1/100–103 lines
cm:	15
kV:	60
mA:	300
Time:	0.33 sec
mAs:	100

6. Place the phantom in the lateral skull position and make an exposure employing the above factors. Develop the film and inspect it under standard illumination. If the film is overexposed (see Figure 9-1A), reduce the mAs by one-half and repeat (see Figure 9-1B). If the repeat film is underexposed, repeat, again choosing between 100 and 50 mAs (70 or 75 mAs) (see Figure 9-1C).

If the final radiograph meets our standards we now have a working technique as follows:

Distance: 100 cm

Region: skull   mA: 300   mAs: 75

Position: lateral   Time: 0.25 sec   Bucky: yes

cm:	13	14	15	16	17	18	19	20	21	22	23
kV:	56	58	60	62	64	66	68	70	72	74	76

Table 9-2. Variable-kilovoltage technique[a]

Grid 10:1/100–103
Medium-speed screens and film
Single-phase equipment

Part	Projection	mAs	SID (inches)	Grid														
Cranium	AP, PA, oblique and lateral	60	40	Yes ⎫	cm	12	14	16	18	20	22	24	26					
					kV	60	64	68	72	76	80	84	88					
Sinuses	PA, frontal	50	36	Yes														
	Maxillary	40	36	Yes ⎭														
Chest	PA	10	72	Yes ⎱	cm	14	16	18	20	22	24	26	28	30				
	Lateral	40	72	Yes ⎰	kV	64	68	72	76	80	84	88	92	96				
Cervical spine	AP, oblique	30	40	Yes ⎫	cm	12	14	16	18	20	22	24	26	28	30	32	34	
	Lateral	100	72	Yes	kV	60	64	68	72	76	80	84	88	92	96	100	104	
Thoracic spine	AP, oblique	30	40	Yes														
	Lateral	50	40	Yes														
Lumbar spine	AP	100	40	Yes														
	Lateral	200	40	Yes														
Pelvis	AP	100	40	Yes ⎭														
Extremities	All	50	40	No ⎬	cm	1	2	3	4	5	6	7	8	9	10	12	14	16
					kV	32	34	36	38	40	42	44	46	48	50	54	58	62

[a] 6 kV added to grid techniques.

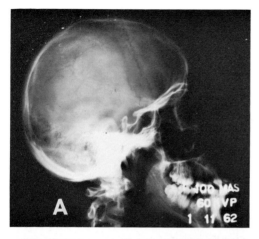

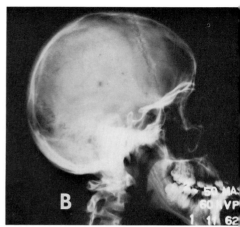

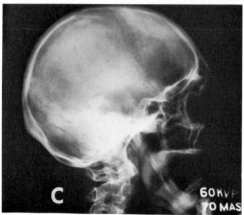

Figure 9-1. Lateral views of a skull. **A.** The Initial film, an overexposure. In **B,** the mAs was reduced to half to compensate, and the result is an underexposure. In **C,** the mAs was adjusted to give a correct exposure by using the average mAs for the two preceding mAs values.

The next step is to apply our mAs correction factors for the adult patients who are above or below normal. This correction is based on a 30 percent change in mAs. In other words, we would add 30 percent to our base or normal mAs for patients who are above normal penetration (Class A), and decrease our mAs by 30 percent for patients who are below normal penetration (Class C).

For uncooperative patients, simply select higher mA values and faster times, maintaining a constant mAs.

Extremity group

Let us take the knee to start this group. Select the mAs you want to use. Suppose, in this case we want to use 10 mAs. We have a choice of mA and time values:

    100 mA at .1 sec = 10 mAs
    200 mA at .05 sec = 10 mAs
    300 mA at .033 sec = 10 mAs

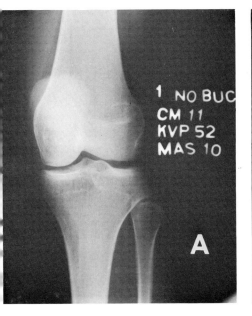

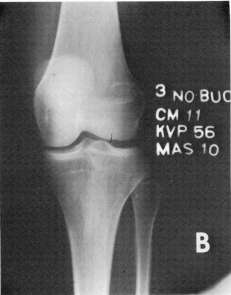

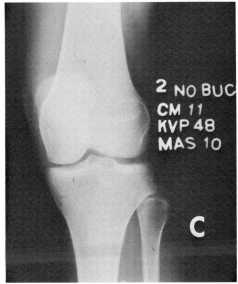

Figure 9-2. Three radiographs of the knee, all at 10 mAs. View **A** is taken at 52 kV, **B** at 56 kV, and **C** at 48 kV.

Since the 100 mA station is on the small focus, let us select 100 mA at 100 ms, a 100-cm SID, and medium-speed screens.

Measure the knee in the AP projection and add 30. In this case assume that the knee measures 11 cm; then, by the formula kV = 2 × cm + 30, we have 2 × 11 = 22, and 22 + 30 = 52. This is our base kV.

Let us expose the knee at 52 kV (Figure 9-2A), make another exposure at 4 kV more (Figure 9-2B), and another at 4 kV less (Figure 9-2C). Process all three films and select the one which meets acceptable standards.

Suppose we select film number 1 (Figure 9-2A). We can now insert 52 kV under 11 cm on our scale.

If more contrast and better detail are desired, a Bucky technique can be used. In this case appropriate grid conversion can be used to compensate for the increased absorption of the grid.

Extremity group: alternative method

Let us select the lateral foot and ankle and an mAs factor that does not work out on the first exposure. To find our kV we would use 2 × the cm thickness of the part (2 × 7 = 14) plus 30, which gives 44 kV. Our technique factors then are these:

cm: 7
kV: 44
Distance: 40 in. (100 cm)
mA: 100
Time: 30 ms
mAs: 5

Cone: to film
Film: regular
Screens: medium-speed
               calcium tungstate
Bucky: no

Expose the part, using the above factors (see Figure 9-3: left view). Upon inspection of the finished radiograph we find we have penetrated the part, but have insufficient density. Expose a second film with an increase in mAs, since mAs is the controlling factor for density. The second radiograph, exposed at 10 mAs (see Figure 9-3: right view), is of the desired quality. Figures 9-4 (left and right views) show the same procedure carried out with dried bones of the foot:

cm: 7
kV: 44
Distance: 100 cm
mA: 100
Time: 100 ms
mAs: 10

Cone: to film
Film: regular
Screens: medium-speed
               calcium tungstate
Bucky: no

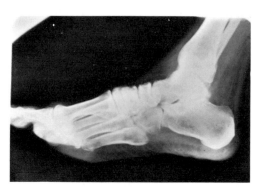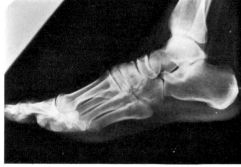

Figure 9-3. At left, lateral view of foot and ankle at 5 mAs. At right, same view at 10 mAs.

By adding 2 kV per centimeter thickness to the base kilovoltage of 30, our extremity chart would be:

mA: 100   Time: 100 ms   Distance: 100 cm   Screens   Bucky: no   Cone: to part

cm:	2	3	4	5	6	7	8	9	10	11	12	13	14	15
kV:	34	36	38	40	42	44	46	48	50	52	54	56	58	60

To compensate for children and infants, we would use the age correction table. (Refer to Chapter 7, p. 181) where the age correction table appears.)

Because of the numerous variable factors in radiography it is often necessary to make two or three exposures of a part to secure the best results. After processing the film under standard conditions, one should study the radiograph on a standard illuminator; if underexposed, the mAs should be increased for the next exposure; if dense and black from overexposure, reduce the mAs by one-half. If gray and flat from overpenetration, reduce the kV by 10 to 15 kV; if it is "chalky," increase the kV by 10 to 15 kV. Further fine adjustments in kV or mAs can then be made in order to give you the type of radiograph you wish.

Effective kV

Effective kV refers principally to the lowest kV with which a part can be penetrated. This minimum penetrability has almost as many variations as there are different types of machines and tubes. Some machines or tubes are more effective at a given kV than others, so it is impractical to set up an mAs and kV balance and regard it as standard the world over.

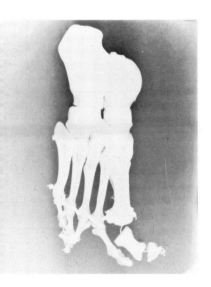
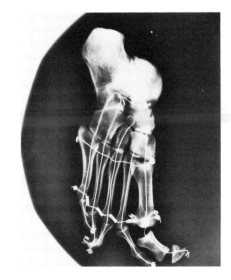

Figure 9-4. Views of bones of the foot (dried): at left, at 40 kV and 1.2 mAs; at right, at 40 kV and 2.4 mAs.

DIRECT EXPOSURE

This particular type of technique is restricted in application to the small parts of the body, and is further restricted by the thickness of the parts themselves. This limitation can be explained by the fact that a normal amount of contrast and detail is necessary for bone radiography and (beyond a certain thickness of tissue) to achieve this is a hopeless task with direct exposure, i.e., cardboard or plastic holder with non-screen or regular film.

There are times when one will encounter conditions which make direct exposures impractical. These instances are usually:

1. Necessity for increased contrast of parts thicker than 12 cm;
2. Necessity for increased contrast of thick muscular parts;
3. Presence of a plaster cast;
4. Necessity for increased speed of exposure.

When these or comparable conditions present themselves, it is advisable to use screens. When screens are used, one must compensate for the increase in speed by reducing mAs. For the proper reduction in mAs, one must know the speed ratio between screens and direct exposure. This will depend, of course, on the kind of screens being used. For example the ratio for medium-speed screens to direct exposure (employing non-screen film) is 5-to-1; therefore, if we have been using 50 mAs for non-screen work, we would use 10 mAs with screens, all other factors remaining the same. If we were using screens and 10 mAs and went to non-screen film, we would use 50 mAs, all other factors remaining the same except processing. Some non-screen film can be processed only by manual processing; thus, it is not to be used with automatic processing.

CLOSE SUBJECT—SID TECHNIQUE

Under certain conditions a short SID, i.e., focal spot 75 cm (30 in.) from the subject being radiographed, will be of value (Figure 9-5).

The short distance takes advantage of the wide divergence of the x-rays. The close position of the tube or focal spot causes blurring of the superimposed parts distant from the film, while the part to be radiographed, being in close contact with the film, still remains sharply outlined. This technique can be used in radiography of the sternum in the PA oblique position, and the temporomandibular joint.

CHARTING RADIOGRAPHIC EXPOSURE

It may be desirable to formulate an exposure chart by charting exposure on a graph and then transferring the desired technique to a permanent place in

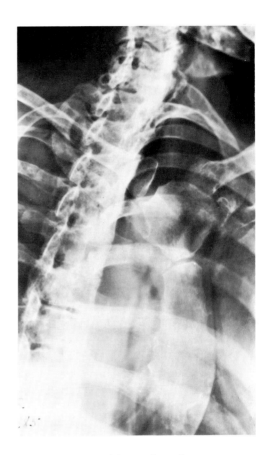

Figure 9-5. Close subject-tube distance technique: PA view of a sternum at 50 mAs and 65 kV, distance 30 in., collimate to part, 8-to-1 grid, medium screens.

the control stand. It is best to have one variable factor and keep the others constant. Either the voltage or the time of exposure (mAs) is most often varied.

As an illustration, suppose a chart is to be made for the lateral lumbar-sacral spine. We should first refer to the capacity of the x-ray unit and also the tube rating chart.

Let us select the following factors:

SID: 100 cm
Screens: medium speed
mA: 100
Collimate: to part
kV: 85

For the chart, take a sheet of graph paper. Mark along the left margin the exposure time in seconds. Along the bottom of the chart, mark the thicknesses of the patient in centimeters. When the first patient is radiographed, measure the thickness of the lumbar spine in the lateral position, select the exposure time thought to be most appropriate to the thickness, and make a dot on the chart at the intersection of the exposure time and thickness. Process the film under standard processing procedure, study the film for detail and contrast with the illuminator which is to be used for interpretation. If the radiograph thus pro-

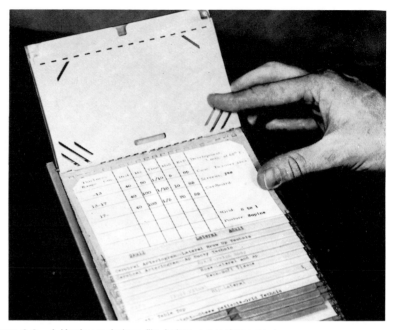

Figure 9-6. A Kardex technique file; below, a specimen card.

Table 9-3. Typical Kardex card for a lateral skull exposure

Thickness range: cms	SID	mAs	mA	Time	mAs	kV	Legend
Infant —							F-Direct exposure holder
Child —							S-Screens, average speed
8–12	40 in.	10	100	$\frac{1}{10}$	10	85	G–Grid
12–17	40 in.	15	100	$\frac{3}{20}$	15	85	
17–21	40 in.	20	100	$\frac{1}{5}$	20	85	
—							
SKULL		LATERAL			Film size 10 × 12 in.		Projection no. 15

duced is found to be satisfactory, the dot may be checked; if found to be overexposed, the dot may be replaced by a plus sign; if underexposed, by a minus sign.

This procedure should be followed until several entries are made on the chart. Some of these will indicate underexposures and some overexposures, but most of them will show correct exposures along a curved line upward, the exposure increasing with an increase in part thickness.

This method may be employed for special procedures or for any technique when charting exposures in which one factor, either the time or voltage, is varied according to the centimeter thickness of the part being radiographed.

ARRANGEMENTS OF TECHNIQUE GUIDES

Technique guides should be posted in the control area of every exposure room. These guides may be in the format of a large chart or as a Kardex. Figure 9-6 and Tables 9-3 and 9-4 show a typical Kardex arrangement. In any event, the technique guides should all be alike so that there is a consistency from room to room. Then a technologist does not have to go through an entire guide just to find suitable techniques—for example, a lateral view of the ankle.

Table 9-4.   Arrangement of Kardex file

Chest	PA	Pelvis	AP
	Oblique		Lateral
	Lateral		
	Stretcher	Hip	AP
	Table Bucky		Lateral
	PA-grid-upright		
	Oblique-grid-upright	Gallbladder	PA
	Lateral-grid-upright		Upright
			Decubitus
Skull	Lateral		
	AP-PA-Waters	Colon	Regular
	Brow up		Air-contrast
	Petrous		Chassard-Lapine
	Mentovertex		
	Mastoid-lateral	Pelvimetry	AP
	Stenvers-PA		AP inlet
	Optic-PA		Lateral
	Mandible		Placenta
	Mandible-lateral		
		Extremities–non-screen	Hand
Sinuses	Lateral		Forearm
	Frontal		Elbow
	Sphenoid-maxillary		Humerus
			Shoulder
Spine	Cervical-AP		Foot
	Cervical-lateral		Os calcis
	Cervical-oblique		Tibia-fibula
	Neck, soft-tissue		Knee
	Dorsal-AP		
	Lumbar-AP	Extremities-screens	Adult
	Lumbar-oblique		Child
	Lumbar-lateral		Infant
	Lumbar-spot, L5		
		G.I. Series	PA-oblique

10 THE PATIENT PROBLEM: PATHOPHYSIOLOGICAL CHANGES RELATED TO TECHNIQUE

Up to this point we have considered the patient only in relationship to size. Now consideration must be given to the effect that the presence of a pathological process or deviation from a normal situation might have on the choice of technique exposure factors.

THE PATIENT PROBLEM

When the patient is being readied for radiography, the physical characteristics and region or part to be examined must be carefully observed. Any information regarding the pathology or condition of the region or regions to be examined will usually prove helpful. The length of time a body part can be held in a comfortable, steady position should be noted.

Experience will enable the radiologic technologist to judge the location of body organs by the physique of the patient, especially the organs of the abdomen, and particularly the gallbladder, stomach, and colon. The technologist should understand the four distinct subject types of body habitus (Figure 10-1)—hypersthenic, sthenic, asthenic, and hyposthenic—since these body types affect organ position. In skull radiography, one should understand the three types of skulls—brachycephalic, dolichocephalic, and mesocephalic (Figure 10-2)—again because the location of internal structures is related to skull types. Depending on their shape, atypical skulls will require more or less rotation of the head or an increase or decrease in the angulation of the central ray. In the brachycephalic skull, which is short from front to back, broad from side to side, and shallow from vertex to base, the internal structures are higher with reference to the baseline, and the long axis is more frontal in position. In the dolichocephalic skull, which is long from front to back, narrow from side to side, and deep from vertex to base, the internal structures are lower with reference to the baseline, and the long axis is less frontal in position. The mesocephalic or so-called normal skull is more or less oval in shape, being wider behind than in the front. The average skull measures approximately 15 cm in the lateral position, 19 cm in the anteroposterior position, and 22 cm in the mentovertex position.

Asymmetry of a skull should also be considered, but if the technologist will

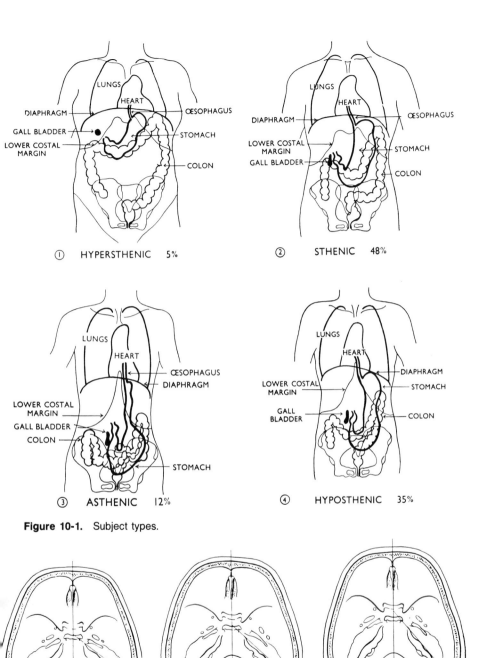

Figure 10-1. Subject types.

1. HYPERSTHENIC 5%
2. STHENIC 48%
3. ASTHENIC 12%
4. HYPOSTHENIC 35%

Brachycephalic   Mesocephalic   Dolichocephalic

Figure 10-2. The three types of skulls.

adhere to the fundamental radiographic baselines and rules of positioning, little difficulty will be encountered. Care and precision should always be exercised, and the use of a protractor or cardboard angles will be tremendously advantageous in producing standard results.

Skulls vary in size and shape, so there is a variation in the position and relationship of internal parts. Internal deviations from the normal are usually indicated by the external deviations; they can thus be estimated with a reasonable degree of accuracy. There is a 4-cm difference between the lateral and antero-posterior measurements of the normally shaped skull. Any deviation from this relationship indicates a change in relationship of the internal structures which, if more than a 5° change, must be compensated for by either a change in rotation of the part or of the angulation of the central ray.

THE FACTOR OF MOVEMENT

Radiographic examinations—particularly of infants and children, emergency radiography, bedside examinations, and all special procedures—must be based on the ability of the patient to hold still and to suspend respiration, and on the ability of the x-ray technologist to successfully immobilize the patient. A sense of how long the patient or part will remain still can be developed. One can also develop the ability to immobilize the part in such a way as to preclude the possibility of motion.

For radiography, motion may be classified into two groups: (a) physiological or involuntary, and (b) accidental or voluntary. Physiological motion is ordinary respiration, heart action, peristalsis, spasm, or tremor. Accidental motion refers to general body movement. Of these two forms of motion, the latter may be more readily controlled. The former can be controlled by employing high mA and short exposure times, or conventional mA and short exposure times in combination with rare earth screens.

For results of the highest quality one should know which of the just-mentioned causes of motion should be considered for each part. For example, radiography of the chest does not require that the time of exposure be based on peristalsis of the kind found in the stomach and intestines. The three most common causes of motion in radiography of the chest are respiration, heart action, and accidental motion. Heart movement and vascular pulsation are generally considered the fastest motion and the most difficult to control. By controlling the fastest motion, any slower motion will also be controlled. If it were not for the problem of motion due to the heart, the time of exposure could be based either on how long the individual can hold still or on the length of time the breath can be held. The normal cardiac cycle being approximately 70 beats per minute, one should try to utilize exposure factors of 16 ms ($\frac{1}{60}$ sec).

In the region of the stomach, peristalsis and diaphragmatic movement—principally peristalsis—are the usual forms of motion to control. Generally speaking, exposure values for this part range from 50 to 100 ms ($\frac{1}{20}$–$\frac{1}{10}$ sec).

With certain types of injuries, muscle spasm, or tremors, the possibility of successful immobilization is minimal. In such cases, the time of exposure must be shortened as much as practical. For this group, the preference is with fixed-kV or high-kV techniques that make use of higher than usual kV and short exposure.

POSITIONING AND MEASURING

Proper positioning (Figure 10-3) of the patient guarantees that desired anatomical structures will be projected into the radiographic field of interest, while proper measuring of these structures determines the x-ray energies that allow the best diagnostic evaluation. Duplicate film densities are possible only when all technologists position and measure the anatomical structures under consideration in the same manner.

Generally, the measurement in centimeter depth for any particular structures is along the line of the centrally projected x-ray beam. Ideally, the part to be radiographed should always be measured by calipers. Do not simply measure to cassette or tabletop; do not make the mistake of computing the measurement of an AP lumbar spine by measuring the depth to tabletop when the spine does not rest on the table.

Sometimes certain anatomical structures present a variable contour. In an AP

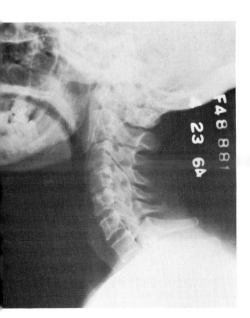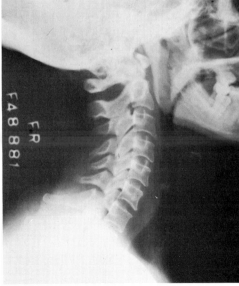

Figure 10-3. Radiograph of the cervical spine showing, at left, improper positioning of the patient and motion; and, at right, the patient in a correct position.

chest radiograph, for instance, an average between the normal measurement and that of the deepest part to be projected into the active field must be considered. In other instances, e.g., the AP lumbar spine, the measurement is indicated at a point 6.25 cm (2.5 in.) below the sternum at the lower margin of the anterior ribs, while the central ray is localized over the umbilicus. These exceptions must be carefully studied in order to guarantee satisfactory end results.

The application of compression bands may cause a considerable change in thickness of the part to be radiographed. Measurements should also be made after the patient has been placed in position for the exposure.

TISSUE ABSORPTION

The x-ray absorption characteristics of tissues are unpredictable, for it is difficult to estimate the physiologic or pathologic state of tissue even with some measure of experience. It is not practical to try to establish these variables on a precise mathematical basis. If the exposure system provides enough latitude for possible errors, and the same body part is being considered in a given projection, a fairly close approximation may be attained in the majority of cases. Small deviations from the normal in tissue density or thickness can be ignored, but compensation must be made for recognizable abnormal conditions of the body by adjustment of the exposure factor selected.

Classification of tissue

By classification of tissue, we mean the relative radiolucency or radiopacity of different types of tissue—emaciation, superfluous flesh, muscle, and normal flesh. In bone radiography, atrophy from age, disuse, or disease must be considered. In examinations of the thorax the type of patient, age, amount and consistency of tissue, and amount of air in the lungs are determined by the structural type of the individual. The consistent production of high-quality radiographs would be comparatively uncomplicated if all tissue had identical opacity to x-radiation. A technologist could rely on a caliper to measure the centimeter thickness of the part to be radiographed and then compensate in exposure factors according to a predetermined scale. But patients cannot be classified by thickness of part alone; type of tissue and pathology are important factors in setting up the proper exposure techniques.

In a general way, patients may be classified as easy to penetrate, normal, or hard to penetrate. Easy to penetrate are the young, the aged, the underdeveloped, and those suffering destructive pathology. Normal are those of medium age and average development. Hard to penetrate are those with muscular builds and those suffering "additive pathology."

To compensate for tissue differences in the small, medium, and large patient, one may deal with three classes. Class A would include thin and emaciated patients, class B, normal or medium patients, and class C, large or muscular patients. For A, B, and C classifications, mAs values could be stated; for example, in the lateral radiography of the three types of lumbar spine, each measuring the same thickness, the A patient would require less mAs and the C patient would require more mAs than the B patient. It is apparent that if the same peak kV and the same mAs were given to all three patients, the A patient's radiograph would be overexposed while the C patient's would be underexposed.

Thickness of part

A variation in thickness of extremities should not cause a great deal of difficulty; it is easier to compensate for variations here than in working with the heavier parts. If the object being radiographed is an ankle swollen to the size of the average knee, then the knee technique should be used; if the individual is of such size that the normal ankle is the size of the average knee, the same holds true. The radiologic technologist with limited experience should spend considerable time in studying the thicknesses of the part which is being radiographed. As experience grows, errors due to poor judgment of the thickness of the subject will become fewer. Variations in thickness and size of the pelvis, abdomen, spine, etc., are much more difficult to judge even for the experienced technologist. When a lack of uniformity exists in the routine radiographs produced in the department, it is usually with the heavier parts of the body. When the subject to be x-rayed is unusually large, there is a tendency on the part of the technologist to use too much penetration and to overexpose the film.

A film which is too light in radiographic density, but which has received full development, or a film too dark in radiographic density after full development, at once tells the radiologic technologist that the technique used has been either too little or too much. In either case it is simple either to increase or to decrease the technical factors for the next radiograph. The experience gained from such a procedure will be far more valuable when the next heavy patient is to be radiographed than the experience gained from employing inappropriate exposure factors and/or improper processing.

Every technologist needs to realize that the thickness of a body part is not always an index of its physiology or its x-ray absorption qualities and cannot be employed by itself as an authoritative index of the radiographic density likely to be produced.

THE CHEST

The chest is less dense to x-rays than any other part of the body of the same thickness. The soft tissues and bones of the thoracic wall are like similar structures elsewhere, but air in the lungs reduces the average density to a marked degree.

Children and infants. Adult chests, though they vary in thickness, are of approximately the same radiographic density. But if a given thickness of the thorax of a child is compared with a similar thickness of the thorax of an adult, the average density to x-rays of that of the child is considerably greater than that of the adult. The difference is most marked in infants and gradually decreases as age advances. This greater opacity makes necessary a special exposure technique for films of the chest of infants and children.

Types of thoracic cavities

Radiography of the thoracic cavity and lungs requires more attention and care than that of any other body group. Results will depend largely on classification of patients into proper groups. The judgment of the technologist in classifying is of great importance.

Heavy-chest type. In this class fall individuals having a heavy, dense muscular frame. Ordinarily, it calls for a plus percentage correction factor, depending on tissue density. Women with very heavy frames and very large breasts would be placed in this group.

Thin-chest type. Such patients are generally recognized, when they are properly positioned against the image receptor with their shoulders and arms thrown forward, by their protruding scapulae and lack of flesh. A minus percentage correction factor is necessary. (Some average patients of normal build will be found with scapulae extending beyond the normal measure.)

Very thin chest type. This individual is the barrel-chested type with less than normal flesh, generally having good expansion, where the thoracic cavity represents practically the entire depth of the thorax. Such patients require a decrease in technique which is accomplished by decreasing mAs below that required for a normal chest. Kilovoltage is not changed, since this would change contrast levels.

PATHOLOGICAL CHANGES

Table 10-1 illustrates common pathological entities seen in radiography. These disease entities can be classified according to corrective changes that must be made in technique factors in order to obtain diagnostic films, that is, as either additive (hard to penetrate) or destructive (easy to penetrate). As the words indicate, additive disease processes require an increase in the technical factors, and destructive, a decrease. Since penetration is controlled primarily by kV, technical factor changes should be made primarily by use of kV.

Table 10-1. Penetrability to x-rays of pathology of body systems, for far advanced disease states

SKELETAL SYSTEM	RESPIRATORY SYSTEM
Additive (hard to penetrate)	Additive (hard to penetrate)
Acromegaly	Actinomycosis
Acute kyphosis	Arrested tuberculosis (calcification)
Callus	Atelectasis
Charcot joint	Bronchiectasis
Chronic osteomyelitis (healed)	Edema
Exostosis	Empyema
Hydrocephalus	Encapsulated abscess
Marble-bone	Hydropneumothorax
Metastasis (osteosclerotic)	Malignancy
Osteochondroma	Miliary tuberculosis
Osteoma	Pleural effusion
Paget's disease	Pneumoconiosis
Proliferative arthritis	Anthracosis
Sclerosis	Asbestosis
	Calcinosis
Destructive (easy to penetrate)	Siderosis
Active osteomyelitis	Silicosis
Active tuberculosis	Pneumonia
Aseptic necrosis	Syphilis
Atrophy—disease or disuse	Thoracoplasty
Blastomycosis	
Carcinoma	Destructive (easy to penetrate)
Coccidioidomycosis	Early lung abscess
Degenerative arthritis	Emphysema
Ewing's tumor (children)	Pneumothorax
Fibrosarcoma	
Giant cell sarcoma	CIRCULATORY SYSTEM
Gout	
Hemangioma	Additive (hard to penetrate)
Hodgkin's disease	Aortic aneurysm
Hyperparathyroidism	Ascites
Leprosy	Cirrhosis of liver
Metastasis (osteolytic)	Enlarged heart
Multiple myeloma	
Neuroblastoma	SOFT TISSUE
New bone (fibrosis)	
Osteitis fibrosa cystica	Additive (hard to penetrate)
Osteoporosis/osteomalacia	Edema
Radiation necrosis	Destructive (easy to penetrate)
Solitary myeloma	Emaciation

Although it is quite common to see test questions on how much kV or mA is to be added or subtracted for a disease entity such as emphysema, in the real world it is next to impossible to determine from looking at the patient or from the radiographic request the exact extent of the pathological process. Emphysema is a good example. Before the radiologist sees any radiological evidence of emphysema—such as irregular radiolucency of the lung fields, increased retrocardiac and retrosternal air space, flattening of the hemidiaphragm and blunting of the costophrenic angles—the patient may have laboratory-demonstrable emphysema. Laboratory pulmonary-function tests will indicate pulmonary emphysema far earlier than a radiologist can see it on the film. When the patient is referred to the radiologist for a chest examination with a medical history indicating emphysema, there are many technologists who automatically decrease technique factors. The unfortunate thing is that the films are almost invariably underexposed and so must be repeated. Hence, even though a disease process may be indicated in the patient's history, the extent of the disease rather than its mere presence should regulate changes in the technical factors.

The same thing holds true for metastasis, particularly bone metastasis. In this case, unless the technologist has a thorough knowledge of medicine, he will not know whether the metastases are osteoblastic or osteolytic nor will he know their extent.

There is only one really accurate way to determine technique changes for certain disease processes *in the individual patient,* and that is by maintaining exposure files on each patient. In this file, one has immediate access to the exposure factors which are suitable for the patient in question. Another way is to mark on the folder jacket of the patient's film that the patient has, for example, "emphysema"—and whatever technical factor change if any, is required.

Although the above measures are time consuming and may be impractical in very large radiology departments, they can still be useful. The point to remember is that for a patient with a spreading and life-threatening disease process, *the exposure of the patient to radiation becomes secondary to the information obtained from providing good diagnostic films.*

Another way to determine what changes should be made in technique factors is to learn the fundamentals of disease processes that often occur. It is not necessary to learn differential diagnosis for a myriad of diseases, but only to recognize that a substantial disease process is present which will require a technique change. When the technologist is capable of ascertaining such processes, he becomes invaluable to the radiologist. The following illustrates the very basic principles of these processes.

THE SKELETAL SYSTEM

Except in cases of trauma, the radiologist is concerned whether there has been destruction of bone (osteolytic) or the production of bone (osteoblastic). In osteolytic bone changes, there is an actual destruction of the bone or cartilage which manifests itself as radiolucent areas within the bone substance. Good examples of this are osteoporosis (Figure 10-4) and osteolytic metastasis such as may occur with metastases from multiple myeloma (Figure 10-5). It is well known that 50 percent of the bone substance must be lost before a change can be seen on radiographic films. Although the presence of osteolytic disease might be thought to indicate a decrease in the technical factors, in most cases

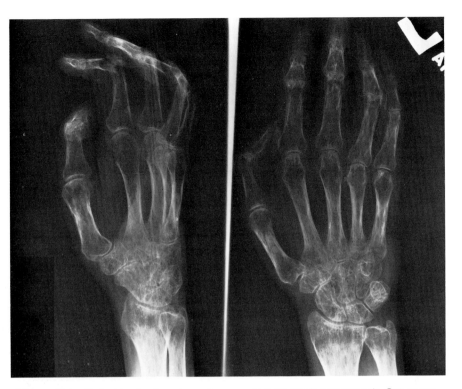

Figure 10-4. Sudeck's atrophy, a form of severe, post-traumatic osteoporosis. Osteoporosis generally requires a decrease in technique.

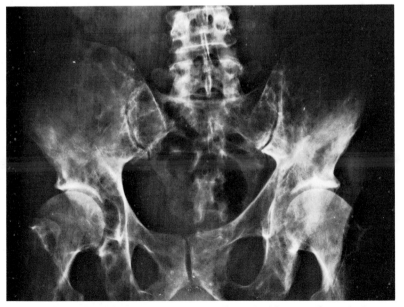

Figure 10-5. Metastases from multiple myeloma. In this case, the metastases represent areas of destroyed bone which appear as small, dark areas on the film, particularly noticeable in the pubic rami.

243

this is not valid, except possibly for widespread, severe osteoporosis (loss of bone calcium). What is far more important is to obtain films with the highest contrast available within practical limits. This can be obtained by changing from standard screens to detail screen exposures, changing the type of film used, or lowering kV. Changing from screen to non-screen exposures will generally lower contrast rather than increase it. The reason for wanting high-contrast films is to allow the radiologist to see early destructive changes which may be masked in films of lower contrast.

The other type of bone change is osteoblastic. Osteoblastic changes indicate new bone or cartilage formation, which generally requires an increase in technical factors, particularly kV. A good example of osteoblastic changes is that from prostatic carcinoma metastasis (Figure 10-6). Osteoblastic changes are seen

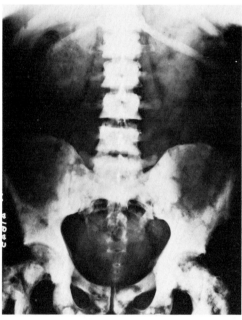

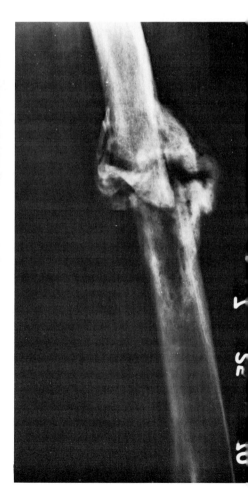

Figure 10-6. Prostatic metastases. An example of osteoblastic metastasis. Note how dense the metastatic lesions (dense white areas) are in comparison with normal bone. Osteoblastic lesions usually require an increase in exposure technique.

Figure 10-7. Chronic osteomyelitis of a fracture in a diabetic patient. The sclerotic appearance of the callus is evident. These lesions require overexposure and high penetration to detect bone activity, so that one can determine healing.

radiographically by an increase in bone density which gives the bone a sclerotic appearance (Figure 10-7). Again, high-contrast films are generally preferred.

Sometimes in association with bone changes, there are soft-tissue changes, edema being the most common. When edema is present, the technical factors must be increased, since the edema presents an overall increase in tissue density. There are also occasions, particularly in early bone tumors, in which both high-contrast and long-latitude films are required—the high contrast for bone detail and the long scale for soft-tissue detail.

As far as fractures are concerned, the initial examinations should be done at the standard routine projection and technique unless the condition of the patient indicates otherwise. In no case should a fracture be moved unless by a physician. In the cases where there is a significant amount of hemorrhage into tissues, technique factors should be increased. Soft-tissue swelling and/or hemorrhage can be detected by gentle palpation of the area in question.

THE RESPIRATORY SYSTEM

The most common disease process seen in the lungs is chronic lung disease, which may or may not be associated with pulmonary emphysema. To detect lung processes requires long-latitude films such as may be obtained with high kV (120kV) and long-latitude films. The use of the high kV also allows better penetration of the mediastinum. To detect bone changes within the thorax requires lower-contrast films. Thus no single film will adequately demonstrate both bony and pulmonary parenchymal structures, and this needs to be clearly held in mind.

Most pulmonary disease processes require standard exposures, as indicated above. From practical experience, if a patient is suspected of having pulmonary emphysema, it seems to be better to make a standard chest film exposure and then cut the mAs in half if the film is overexposed (Figure 10-8). In the author's (TTT) experience this seems to require fewer repeats than automatically decreasing technical factors simply because the patient "has emphysema." For the other disease processes such as pneumoconiosis, bronchiectasis, and metastasis, the extent of the disease dictates the amount of technique change, if any. The radiologist generally wants to see through the disease process and this should dictate what technical factors are chosen. The amount of technical factor conversion is determined only from experience, and there are no hard and fast rules that will work in all cases.

A good example of changing technical factors is a Pancoast tumor. A Pancoast tumor is a bronchogenic carcinoma that arises in the apex of the lung adjacent to the mediastinum (Figure 10-9). A number of these tumors appear as an increased density in the upper lung field adjacent to the mediastinum, which are shown best by using standard exposure techniques. However, by acquiring films which adequately penetrate the tumor, rib destruction may be seen which is almost pathognomonic of this tumor. Hilar lymph node involvement such as may

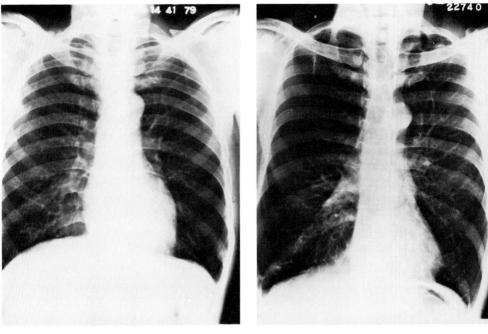

Figure 10-8. Emphysema. The patient on the left has laboratory-demonstrable pulmonary emphysema, but no radiographic changes are noted. No technique change is required. The patient on the right has radiographic-demonstrable pulmonary emphysema, which requires a decrease in exposure. Many laboratory methods detect a disease process earlier than it can be shown radiographically, and one must take into consideration the extent of the disease process as well as the presence of the process.

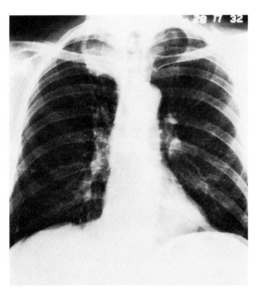

Figure 10-9. Pancoast tumor. Note the density in the right apex adjacent to the mediastinum. The full-size chest film requires no technique change; however, coned films of the lesion may be necessary so that one can ascertain rib destruction.

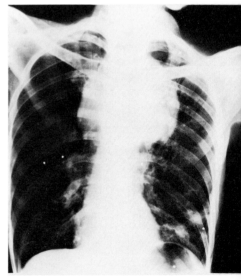

Figure 10-10. Massive hilar and mediastinal adenopathy. Films such as these require no technique change. Care should be taken when utilizing single field phototimers, since phototiming a lesion such as this invariably means overexposing the lung fields and "burning out" small metastatic lesion(s)

246

be seen in sarcoidosis, primary tuberculosis in children, and Hodgkin's disease require no technique change (Figure 10-10).

GASTROINTESTINAL TRACT

There are only two major considerations in regard to the abdomen; bowel obstruction and ascites. In bowel obstruction, a determination must be made as to how much air is trapped in the alimentary tract (Figure 10-11). This can usually be determined by simple palpation of the abdomen. If the abdomen is distended, sometimes even taut, simple palpation will usually show that this distension is caused primarily by air. If the distension feels like dough, it is most likely due to accumulated fluid or ascites. With air distension, technique factors are generally decreased. With fluid distension, technique factors must be increased, particularly kilovoltage. The presence of significant fluid or ascites also means more scattered/secondary radiation fog in the film with its concomitant decrease in contrast. In these cases it is wise to use a high-ratio grid to improve visibility and contrast.

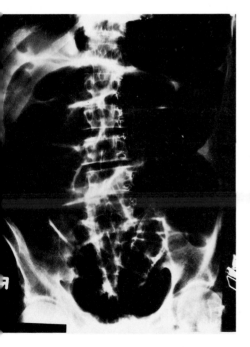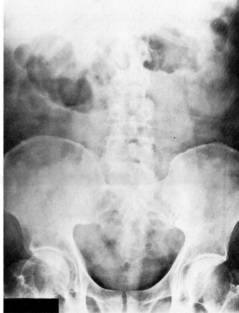

Figure 10-11. Both of these patients have bowel obstruction. The patient on the left has gaseous distension; in the one on the right, fluid predominates. Technical factors would be different for each case.

Cahoon's Formulating X-Ray Techniques

Figures 10-12 through 10-15 are examples of technical changes required for the presence of pathology.

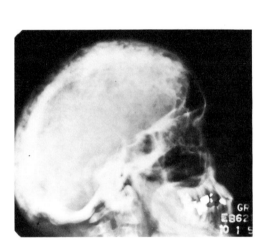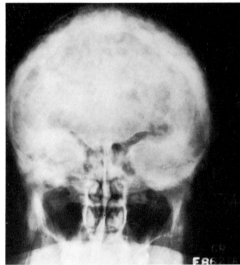

Figure 10-12. At left, plus 40 percent correction factor for Paget's disease. At right, the same correction factor.

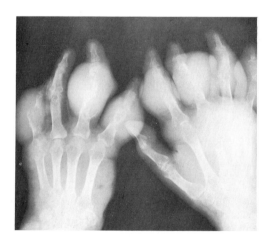

Figure 10-13. Postero-anterior view of hands, with a minus 30 percent correction in mAs for gout.

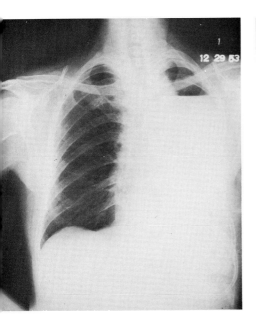 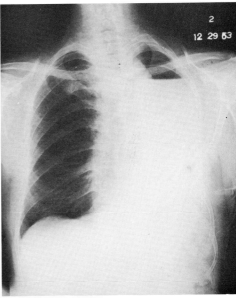

Figure 10-14. At left, chest with fluid, given normal exposure. At right, same chest with the correction of an added 5 kV.

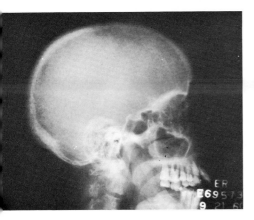 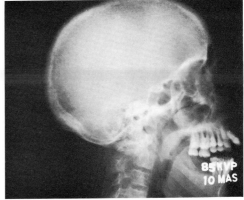

Figure 10-15. At left, lateral view of skull with hyperparathyroidism, with normal exposure. At right, the same skull with a minus 30 percent correction in mAs.

SUGGESTED READING

Felson, B. *Fundamentals of Chest Roentgenology.* Saunders, Philadelphia, 1960.

Felson, B., Weinstein, A. S., and Spitz, H. B. *Principles of Chest Roentgenology.* Saunders, Philadelphia, 1970.

Hipona, F. A., ed. "Cardiac Radiology: Medical Aspects." *Radiol. Clin. North Am.* 9, no. 3 (1971).

Jones, M. D. *Basic Diagnostic Radiology.* Mosby, St. Louis, 1969.

Simon, G. *Principles of Chest X-Ray Diagnosis.* Butterworth, London, 1962.

Squires, L. F. *Fundamentals of Roentgenology.* Harvard, Cambridge University Press, 1964.

Storch, C. B. *Fundamental Aids in Roentgen Diagnosis.* Grune & Stratton, New York, 1951.

Thompson, T. T. *Primer of Clinical Radiology.* Little, Brown & Co., Boston, 1973.

REVIEW QUESTIONS

1. In assessing a patient prior to performing a radiographic examination of the pelvis, the radiologic technologist should determine

   a. Positive identification of the patient
   b. Physicial/mental ability of patient to cooperate
   c. Reason for doing the examination
   d. Last menstrual period for females
   e. All of the above

2. In a brachycephalic skull, the internal structures are in what relation to those in a normal skull?

   a. Higher
   b. Lower
   c. Same relative position

3. In a normally shaped skull, there is normally what difference between the lateral and antero-posterior measurement of the skull?

   a. 1 cm
   b. 2 cm
   c. 4 cm
   d. 8 cm

4. A chest x-ray should be exposed ideally at what exposure time?

   a. 3 ms
   b. 16 ms
   c. 100 ms
   d. 200 ms

5. High kilovoltage techniques require _____ exposure times and _____ radiation exposure to the patient.

   a. Longer, higher
   b. Shorter, higher

    c. Shorter, less

    d. Longer, less

6. Changes for exposure techniques for pathological conditions depend primarily upon

    a. Type of equipment being used

    b. Type of processing

    c. Extent of disease

    d. What the referring physician wants

    e. Extent and type of disease present

For the pathological conditions listed in Column A, select from Column B an appropriate change in technical factors.

Column A	Column B
7. Minimal emphysema	a. Increased exposure
8. Severe emphysema	b. Decreased exposure
9. Sudeck's atrophy	c. No change required
10. Widespread osteolytic metastasis	
11. Acute osteomyelitis	
12. Emaciated patient	
13. Use of compression bands	
14. Gaseous bowel distension	
15. Sthenic patient	
16. Anasarca (generalized edema)	

11 PROCESSING

As suggested in Chapter 2, the processing room is perhaps the most important single room in the radiology department, because the darkroom is the place where the latent image, placed on the film by radiographic equipment, is converted into a useful visible image. It was the opinion of most of those in the radiology profession that the advent of mechanized, or "automatic," radiographic processors would eliminate most of the problems associated with the darkroom. Yet the darkroom continues to be the room in the X-ray department which radiologic technologists and radiologists can least afford to neglect. Radiography begins in the darkroom; this is the place where one ordinarily puts the film in the cassette. Radiography ends in the darkroom; this is the place where the film is processed and made useful for viewing. A technologist can never hope to be a really good technologist unless the darkroom and processing technique are understood and learned. The best and most expensive radiographic equipment is of no avail if one tolerates carelessness or does not follow proper techniques for processing the x-ray film. The processing room can make or break a department as far as quality of the radiographic film is concerned, whether the film is processed manually or mechanically.

THE PROCESSING ROOM

During the past twenty years we have seen a thorough change in the concept of processing radiographic film. The processing of radiographic film by hand (manual processing) has almost been eliminated in favor of mechanized or automatic processors. (The word "automatic" is a misnomer. The process is not altogether automatic. Automatic processing would entail having the film transported from the exposure area of the radiographic equipment to a film processor, and from there emerging as a fully processed film.)

The cost of processors has decreased over a period of time, so that a department can afford to buy several processors. Instead of one central darkroom area, small darkrooms can be dispersed throughout the department, and on occasion even be located in surgery units or elsewhere in the hospital for convenience in doing bedside radiography. With the development of automated equipment such as the Picker Rapido System * or the GE Change-X Chest Film Changer,† the processor can be installed in tandem with the equipment, and no darkroom is required except for loading film magazines. Here, film processing is truly automatic. Whether the darkroom is of centralized type or dispersed throughout the department there are some fundamentals that one needs to understand.

Light-tight entrance

An entrance which provides an easy access and complete protection from outside white light is essential to an x-ray processing room. In a smaller, dispersed darkroom, the room can be made light-tight by simply adding weather stripping around the edges and bottom of the door. Unfortunately, with such an arrangement, one cannot go into the darkroom while someone else is processing film or loading and unloading cassettes. In larger darkrooms, a light-tight entrance can be made by means of two doors and a vestibule arrangement, in which one door is shut before the other is opened, so no light can enter the darkroom itself. Another and similar mechanism is an arrangement of offsetting walls, with the corridor painted black. The black paint absorbs light, and no white light can enter the darkroom. Since in the radiology department everyone always seems to be in a hurry, it would be wise in the system of two doors to have them electrically interlocked so that one door cannot be opened while the other is open. An entire bin of radiographic film can be ruined by white light exposure if the darkroom door is opened while the bin is open. Thus, several hundreds of dollars of film can be ruined if one goes through the two doors so quickly that white light penetrates the room.

Ventilation

The darkroom must be well ventilated, perhaps more than any other place in the radiology department. Besides maintaining humidity, which is necessary to control the development of static electricity, safety standards require that fumes that develop from either manual or mechanical processing be vented from the room. Mechanical or automatic processing does not eliminate the problem of humidity and chemical odor; this depends entirely upon how the unit is installed, and where the silver recovery and replenishment tanks are located. In addition to humidity, the temperature must be controlled, both for physical comfort and for film storage and handling. If humidity and temperature are high, the operators tend to perspire and their damp fingers are liable to leave finger marks on films and screens.

Decoration

It is neither necessary from the technical standpoint, nor desirable from the physiological one, that the processing room be drab and all-black; such a decor

* Picker Corporation, Cleveland, Ohio.
† General Electric Company, Milwaukee, Wisconsin.

should be avoided. The darkroom can be painted almost any pastel or semigloss color. The light level in the darkroom is really controlled by the light intensity and hue emanating from the safelight lamp; it has little to do with the color the walls are painted.

The color of the top of the loading bench should be chosen with the thought that film, paper, and other objects must be readily distinguished. One of the most durable materials used for a loading bench top is sheet plastic or Formica. The loading bench must be well grounded to prevent the development of static electricity.

Safelight illumination

Most darkrooms use a Kodak Wratten 6-B safelamp filter * for conventional x-ray film. With this safety lamp it is possible to see clearly in the processing room and not expose radiographic film. When fitted with a 7½- to 15-watt bulb, exposed film can be left exposed to safelight illumination without fogging for one minute at 3 feet from the safelight or for half a minute at 2 feet. One must remember that the inverse square law holds for light just as it does for x-radiation.

Exposed radiographic film is more sensitive to safelight illumination than unexposed film (Figure 11-1). This increased sensitivity of film to exposure to safelight illumination is called *latensification*. In most cases, x-ray film is twice as sensitive to a safelight exposure after being exposed radiographically. The effect on previously exposed x-ray film of an exposure to a safelight is known as *post-exposure fog*. The quality of the radiograph may be impaired by unnecessary exposure to safelight illumination. This may be a reason for variations in densities encountered in routine radiographs and for a lack of brilliancy that is frequently attributed to scattered irradiation (Figures 11-1 and 11-2). Safelights generally expose only one side of the two-emulsion film, and the amount of fog depends on the amount

* Eastman Kodak Company, Rochester, New York.

Figure 11-1. A photographic example of postexposure fog.

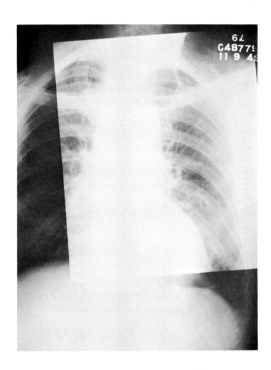

Figure 11-2. Postexposure fog. The lighter portion of this radiograph was protected by a sheet of cardboard during the safelight test which fogged the rest of the film.

of exposure that has been made of the film. For a radiograph with a low overall density, such as a chest radiograph, an increase in the safelight fog brings an overall loss of detail; in a heavily exposed film such as a film of the thoracic spine, a slight increase in safelight fog is less detrimental.

The Kodak Wratten 6-B safelight filter has been recommended and used in darkrooms for years. With the introduction of orthochromatic-sensitive film, the safelight filter has to be shifted more to the red portion of the visible spectrum. To achieve satisfactory safe working times in the darkroom with this green sensitive film, Kodak has developed a safelight filter called a GBX.

It must be emphasized that the 6-B safelight filter is not suitable for use with orthochromatic-sensitive films. If an orthochromatic-sensitive film is exposed to a Wratten 6-B filter, the image will show an apparent increase in speed (low latensification), a very noticeable decrease in contrast, and a significant increase in gross fog. If green-sensitive film is exposed to a safelight filter such as a GBX, there is no significant safelight fogging latensification for up to a 5-minute exposure.

Safelight tests

To test a darkroom for safety from exposure to a safelight, one should place an unexposed film directly on the bench or on the processor under the safelight for periods of 1, 2, 3, 4 and 5 minutes. After each one of these time frames, the film should be processed and compared with an unexposed, processed film.

White-light leaks around doors or windows can be detected by simply turning off all the lights within the darkroom and looking for the telltale signs of white light as it penetrates the edges of the door or window. The safelights above processors, as well as those above the workbench, should be examined. Other equipment located in the darkroom may also have safety lamps associated with it, and one should make certain that these lamps will not expose film.

DARKROOM CLEANLINESS

Owing to the sensitivity of x-ray film emulsions, cleanliness is essential in processing areas. The processing room as well as the accessory equipment must be kept scrupulously clean (Figure 11-3). Spilled solutions should be wiped up at once; otherwise, on evaporation, chemicals may get into the air and later settle on film surfaces, causing spots. Films should be handled with care in the processing room, or artifacts may be created, as will be discussed below in this chapter.

FILM FEEDING

Figure 11-4 shows proper film feeding procedures. The arrows indicate the direction in which film is fed into the processor.

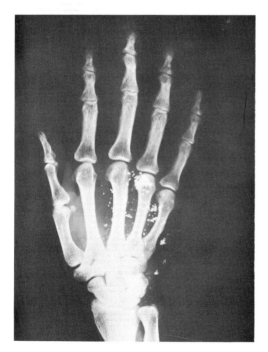

Figure 11-3. Cigarette ashes in the cassette.

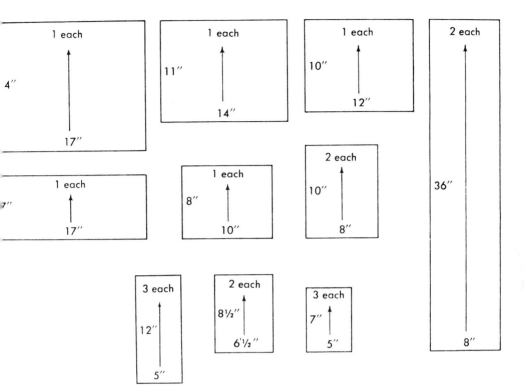

Figure 11-4. Film-feeding guide for mechanical processors.

Figure 11-5. A leader for film feeding of occlusal film or 35 mm copy film.

To avoid overreplenishment, it is advisable to feed all narrower films side by side. Feed films into the processor square with the edge of the side guide. Feed multiple films simultaneously.

For occlusal film or roll film, in order to provide proper transport, use a leader of sheet film (Figure 11-5). The leader should be as wide as or wider than the roll film and at least 7 inches long. Using a 1-inch-wide tape, such as Scotch brand Electrical Tape No. 3, butt-splice the film to the leader, making sure that the adhesive side of the tape is not exposed. Most other types of tape are not suitable, owing to the solubility of their bases in the processing solution.

RADIOGRAPHIC PROCESSING

The function of processing is to convert the latent image into a visible image. Processing is accomplished by systematically moving a film, either manually or mechanically, through a series of solutions and ultimately drying the film. Figure 11-6 shows this process in a mechanical processor. To adequately process a film, the technologist needs a thorough understanding of the chemicals used in processing, their activity, the relationship between time and temperature, and the factors which control processing and on which the system of processing is based. Manual processing for all practical purposes is disappearing from the radiology department, and most of the discussion below will relate to mechanical (automatic) processors.

THE PROCESSING CYCLE

Automatic processors can be subdivided into a number of interdependent systems:

1. The chemical system
2. The transport system
3. The replenishing system
4. The circulation/filtration system
5. The drying system

The chemical system is composed of the developer, fixer, and wash water subsystems. The transport system includes the roller rack assemblies and drive motor. The replenishing system includes the replenishing tanks, replenishing pumps, and related items. Finally, the drying system includes the blower and thermostatically controlled forced air which dries the film. In addition, each of these systems is supported by an electrical system in some manner, but it is outside the scope of this text to delve into the electrical system supporting an automatic processor.

THE CHEMICAL SYSTEM

The activity level of the chemistry in a system is important since one chemistry may be less or greater than another, and one may have been developed for operation at a higher temperature than another. For example, Du Pont XMD chemistry is a high-activity chemistry designed to be used in a short cycle with 90-sec processors. XAD chemistry, on the other hand, is a less active chemistry and is designed for longer processing cycles. Using either of these chemistries at a temperature for which it has not been designed will alter the film response to processing. The more active chemistry will produce a higher density than the less active chemistry at equal temperatures. The activity of the developer must be maintained through proper agitation, replenishment, and circulation of the chemistry, which varies according to the type of processor being used.

Developer

The first step in the chemical system of processing is the developer. The function of the developer is (1) to swell the emulsion to allow the developing agents to diffuse into the gelatin layer more readily and to allow oxidized developer to diffuse out; and (2) to reduce exposed silver halide crystals. The chemistries used in modern automatic processors for development are known as *PQ developers* because the primary agents are phenidone and hydroquinone. (MQ develop-

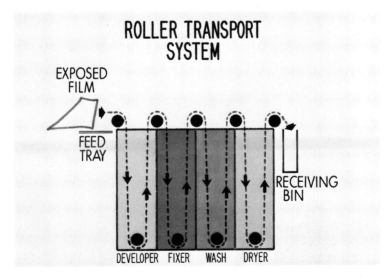

Figure 11-6. Scheme for a mechanical (automatic) processing transport system. An exposed film is fed through a series of paired or offset rollers sequentially through developer, fixer, and wash tanks and then through a dryer chamber.

ers, composed of metol and hydroquinone, are used primarily in manual processing.)

Phenidone acts quickly and is quite sensitive to temperature changes. Its function is to reduce silver bromide to black metallic silver, and it affects mostly the grays of the radiographic film. The phenidone works synergistically with hydroquinone and controls primarily the toe, the speed, and the minimum density of the sensitometric curve.

Hydroquinone on the other hand is slow working and not as sensitive to time and temperature relationships as phenidone. As hydroquinone works synergistically with phenidone, it affects mostly the blacks, and is primarily responsible for the shoulder of the characteristic curve, contrast, and maximum density. It is also the primary buffering agent to keep all the chemicals in balance.

The developer used in automatic processors also includes other chemical elements. One of these is potassium bromide, a restrainer or antifogging agent. Some of the newer antifogging agents are made of organic compounds. Sodium carbonate serves as an activator and provides the alkalinity required for reducing agents to work in. It also assists in swelling the gelatin in the emulsion. Sodium sulfite is a preservative and protects the reducing agents from aerial oxidation. Glutaraldehyde is also added to control swelling of the emulsion and helps to harden it. All the chemicals are dissolved in water, which provides the solvent base and also furthers the swelling of the gelatin.

Starter solution. As time passes, the chemical activity of a solution will gradually fall as it is used up in processing. Fresh chemistry when first mixed is very active. So a synthetic depression of activity is obtained by adding to it *starter solution,* which lowers the activity of the fresh solution to a predetermined stable, uniform level. Starter solution is generally potassium bromide and is a restrainer.

Mixing of developer. Modern developer chemistries are usually supplied in three parts, A, B, and C. Part A is a very strong alkaline material, approximately pH 12. Parts B and C are moderately strong acids and measure a pH 3 (sometimes as strong as a pH 2). The resultant chemical mixture has a pH of approximately 10.6, and a specific gravity of about 1.100.

To mix the developer solution properly, first clean the storage tanks in the processor. Add a volume of water to the replenisher tank mixing vessel according to the directions supplied with the chemistry. Next, slowly add part A and gently stir while adding. After part A has all been added, continue stirring until the concentrate is well blended. Then slowly add part B, and then part C, following the same directions. *Do not use a water hose for stirring. This introduces air bubbles, which oxidize the chemistry.*

Developer oxidation. In radiographic processing the developer plays a number of significant roles. The first of these is development of the visible image; in this the entire silver halide crystal is reduced. The aggregates of these reduced crystals form the visible image. The developer does this selectively by attacking only sensitized silver halide crystals under controlled conditions. The developer also performs a role of amplifier. According to the Guerny-Mott theory, the excitation of the sensitivity speck (as described in Chapter 2) allows the developer to reduce all the silver atoms in the halide crystal, thus increasing the number of reduced silver atoms by over one million times.

The developer, in providing the electrons for the reduction of the silver halide crystals, undergoes an oxidation/reduction reaction. The developer is oxidized by giving up electrons, and the silver halide is reduced by gaining electrons. Thus, the activity of the developer is lowered; as it gives up electrons, it becomes oxidized, simply by being used. Developer can also be oxidized by being exposed to air or by improper mixing of the components of the developer as just noted. Foreign contaminants, such as system cleaners, and contaminating fixer dissolve silver out of the film before development has occurred.

Contaminants of developer interfere primarily with its functions of swelling the emulsion and contributing electrons for reduction. One of the most common contaminants is fixer, which gets into the developing solution usually by cross contamination from the fixer tank into the developer tank. Residues from systems cleaners dissolve silver bromide crystals into the developer solution. Later, the silver ion is reduced to colloidal silver; this is the black sludge sometimes seen if systems cleaners are not adequately removed from the processor after cleaning. Oil applied in lubricating transport racks can accidentally drop into the developer tank and interfere with the reaction by preventing penetration of the emulsion by the developer.

Fixer

Besides developer, the chemistry system includes the fixer. Developer converts the latent image into the visible image, but this process will continue indefinitely and will eventually create a chemical fog unless controlled. The development action is stopped by means of an acid bath called the *fixer*. Fixer is commonly known as *hypo* (sodium thiosulfate). The functions of the fixer are (1) to neutralize the developer, (2) to clear away undeveloped silver halide crystals, and (3) to shrink and harden the gelatin. Modern hypo is composed of ammonium thiosulfate, which forms a complex with silver, producing silver thiosulfate. Fixer dissolves out undeveloped silver bromide. In addition to the ammonium thiosulfate, the fixer contains acetic acid, which fully neutralizes the developer. An aluminum compound is added to shrink the gelatin and harden it.

Mixing fixer. To mix the fixer solution properly, the directions supplied by the manufacturer must be followed. As a general rule, add the proper amount of water to a clean replenisher tank or mixing vessel. Slowly add part A and thoroughly mix the chemicals. Then slowly add part B and stir with a paddle. *Do not use a water hose to mix chemicals.* Part A is a moderate acid, with a pH of 5, and part B is an extremely strong acid, with an average pH of 0.5. The mixed chemicals will have a pH of approximately 4.1 and a specific gravity of 1.110.

Wash water

Washing the film is the last process before the film is dried. The function of wash water is to remove chemicals remaining in the emulsion and those on the surface of the film. The temperature of the wash water is usually 3 centigrade degrees (5 Fahrenheit degrees) lower than that of the developer solution, simply because it is easier to control temperature by heating up than by cooling down. Wash water helps control the stability of the developer temperature.

A problem with wash water is an overgrowth of algae—algae are usually too small to be filtered out by the filtering system. Overgrowth in the wash water may be controlled by the addition of 6 ounces of a liquid laundry bleach such as Clorox, or an algicide, per 2 or 3 gallons of water. Films should not be fed through a processor until all the bleach has been flushed out of the tank. A wash-water circulation of 2 or 3 gallons per minute is usually required for most processors. When the water supply is shut off, the wash tank should be made self-draining to help control algae growth.

Time and temperature relationship in processing radiographic film

There is a relationship between the time of development and the temperature at which development occurs which is based on the specific activity of the developer. If the time of the development is increased, then the temperature of the developer must be decreased, and vice versa. By increasing time or temperature, the contrast and relative speed of the film is increased, and vice versa. Controlling time and temperature is necessary for consistency and sensitometric quality. Table 11-1 shows the relationship between developer temperature and time. Table 11-2 shows the average development time within different types of automatic processors. Note in Table 11-2 that the film immersion time is different for different processors.

Table 11-1. Time/temperature relationships in mechanical processing [a]

Time	Temperature
18 sec	92° F
22	90°
26	88°
30	86°

[a] Figures given are for Du Pont XMD developer. Temperature corrections most likely will be required for other developers that have different chemical activity levels. As a general rule, a 4-sec decrease in developer time requires a 2-degree increase in developer temperature, and vice versa.

There is a time and temperature activity relationship for fixer, just as there is for developer. Normally, fixing time is twice that of *clearing time*. To determine clearing time, take a piece of film, exposed or unexposed, and dip a corner of the film in the fixer until the emulsion clears. This will take about ten seconds. Next, immerse the film further into the fixer and record the time needed for film areas to match. An average clearing time is approximately seven seconds. Agitation again is important and all processors should provide adequate agitation during transport.

DETERMINING FILM DEVELOPMENT TIME

To calculate film development time in a processor, only the time of total immersion in the developer should be considered. To measure this time, turn the processor off. Measure from the entrance and exit roller points to the solution level and mark these distances on the leading edge of a film. This will be approximately 10 cm. Turn the processor on and feed the film through the developer rack. Begin timing when the mark reaches the input roller; stop timing when the film exits the last roller pair. By this method, only the total immersion time is measured.

Another method of calculating the developer timing is to feed a film roller to roller in the developer rack and measure with a stop watch how long it takes the film to go through the rack. Measure the distance from the rollers to the solution at the entrance and exit. Convert to seconds and subtract from the total time to get net developer immersion time. The *average* developing time in a 90-second processor is 20 seconds.

Processing cycles. The standard processing cycle time is 90 seconds. Processors may have cycles ranging from 7 minutes to 3.5 minutes to 90 seconds. The 3.5-minute processor is called the *double-capacity processor.* However, a 90-second processor which has been converted to run at a 3.5-minute cycle is called a *half-cycle processor.* Standard development time in a 7-minute cycle processor is 100 to 120 seconds; in a 3.5-minute processor, 50 seconds; and in a 90-second processor, 20 seconds.

Table 11-2. Characteristics of common processors

| | | Immersion times (sec) | | | |
Processor	Dry-to-drop	Develop	Fix	Wash	Dry
Kodak M6	96 sec	20	16	10	26
Kodak M6AN	93	19	15	9	31
Kodak M7	153	22	17	18	35
PAKO XU	107	20	20	20	21
Picker Diplomat	104	26	19	19	26
General Electric	180	26	26	26	39

Dry-to-drop time. The term dry-to-drop time refers to total processor access time—that is, how long it takes a film to completely traverse a processor. It includes the time required to feed the film into the processor, to transport it through the chemicals into the drying system, and then to deposit it in a drop tray. Dry-to-drop time varies from processor to processor. Not all 90-second processors have a processor access time of 90 seconds. Because of variations within 90-second processors, the more appropriate term is *short-cycle processor.* To determine dry-to-drop time, feed a 14-inch film into the processor and determine the length of time it takes for the film to traverse the processor completely and to drop free into the receiver tray.

Processor temperature control. There are two mechanisms used to control and adjust processing developer temperatures. Gross control of temperatures is effected by a water-mixing valve, a device that interconnects the incoming hot and cold water. The water from the mixing value is fed into a jacket surrounding the developing and fixer tanks. The temperature of this water is 5 Fahrenheit degrees less than the optimized developing temperature.

Fine control of developer temperature occurs within the developer tank itself. It is accomplished with a thermostatically controlled heating element which heats the developer to a prescribed level. Unfortunately, this type of heater is subject to overshooting the prescribed developer temperature, and it is common to see fluctuations of 1 or 2 degrees in developer temperature.

Another method of internal heating is by means of a *heat exchanger.* A heat exchanger is a device through which warm water is passed and heated to a prescribed level. The heat exchanger is located in the bottom of the developing tank in most instances. Heat transfer from the heat exchanger to the adjacent developer fluid maintains developer temperatures at an accurate level.

Display thermometers. Almost all processors have display thermometers which indicate the temperature within the processing tanks. These display thermometers are, more often than not, inaccurate. To determine processing developer and fixer temperatures, a standard non-mercury calibrated glass or bimetallic stem thermometer should be used. An alcohol thermometer is best; if a mercury thermometer breaks, it will contaminate the developer system, and it is next to impossible to remove the mercury from the developer tank.

THE TRANSPORT SYSTEM

The function of the transport system is to move the film through the various components of the processing cycle at a constant speed. The speed at which the film is moved is controlled by a motor drive system. Transport of the film is accomplished by a continuous roller conveyor system (Figure 11-7). By making the roller conveyor system move at a fixed rate, the time each film is in each chemistry is controlled and constant as long as the motor drive system speed is unchanged.

The rollers in the transport system can be either paired (directly across from each other) or offset, and fixed or spring loaded. The rollers are attached to a rigid face or end-plates which provide them with bearing support and maintain this rigidity. The rollers along with the face plates and mechanical supports form what is called a *vertical rack assembly.* There is a vertical rack assembly for the developer, one for the fixer, and one for the wash water.

Two types of rollers are commonly used in an automatic processor. The *planetary roller* is a 1-inch-diameter roller which is the major roller within the processor and provides movement and guidance of the film during transport. Planetary rollers are placed in the vertical rack assembly in such a manner that just enough pressure is exerted on the film to provide transport, but not so much as to cause transport film artifacts such as scratches or pressure marks. The *master,* also called the sun roller or solar roller, is a 3-inch-diameter roller that guides the film around a turn. The master roller occurs mostly in older processors. In some newer processors, the master roller has been replaced by a *semi-master roller* which is 2 inches in diameter and serves the same function as the master roller.

A *turn-around assembly* is the system of planetary and semi-master rollers and appropriate guides located at the bottom of the vertical rack assembly. Its function is to turn the film from the entrance portion of the vertical rack assembly into the exit portion of the rack. The turn-around assembly is not a separate rack, but a component of the vertical rack assembly. There may or may not be a turn-around assembly for the dryer rack, depending on how the film is dried.

TRANSPORT SYSTEM

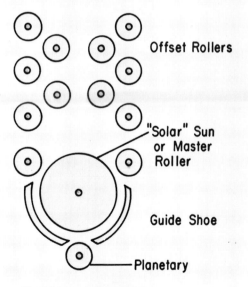

Offset Rollers

"Solar" Sun or Master Roller

Guide Shoe

Planetary

Turnaround Assembly

gure 11-7. Scheme of vertical rack roller trans-
ort system using offset rollers.

The *crossover rack* is a small rack assembly whose function is to feed the film from one rack assembly to another. Most processors in use today have three tanks, and only two crossover rack assemblies. Some newer processors do not have crossover racks.

Guideshoes are present in most processors. These are deflector plates that guide the film into the crossover and turn-around assemblies.

Transport speed is controlled by the speed of the motor drive assembly. In most cases the drive motor runs at 1750 revolutions per minute (rpm), but through a series of reducing gears, the speed is changed to 10–20 rpm. Transport speed may be increased or decreased by manually changing the reduction gears or the drive gears. Variable-speed motor drive assemblies are available and provide exact control over the film immersion times. By increasing or decreasing the motor speed through a rheostat control, the film immersion time can be correspondingly increased or decreased. Thus film development can be controlled by controlling immersion time rather than manipulating temperature.

The motor drive system can be coupled to the transport rack assembly by either direct drive gears or through a series of drive belts.

Racks in the processor should be cleaned with a cleaning pad such as 3M Scotchbrite. Care must be taken not to damage the rollers. Scouring pads loosen dirt so that it can be washed away with water. System cleaners should be discouraged. This is mandatory if the processor has hollow rollers; the cleaner gets into the rollers and contaminates processing solutions. During cleaning, care must be taken that roller springs are not dislodged. The alignment of the guideshoes and rack assemblies should be inspected at the same time and alignment corrected if necessary. Crossover racks should be washed daily; vertical rack assemblies should be cleaned on a periodic schedule, usually every week or two. When removing racks from the processor, take care that fixer is not splashed into the developer.

THE REPLENISHMENT SYSTEM

Proper replenishing is necessary for maintaining a constant level of chemical activity. Replenishing compensates for loss of chemical activity by direct oxidation, aerial oxidation, and contaminants, and for loss of developer or fixer by carryover into the adjacent processing tank and for volume loss by air evaporation. "Carryover" refers to the fact that as a film is developed or fixed, there is always a small volume of chemistry which adheres to the film and is thus moved from one tank to the next.

Replenishing rates

The amount of chemistry that is to be replaced is determined by estimating how much film is used daily. Each manufacturer suggests specific replenishing rates for particular applications. Replenishing rates are normally established on the basis of how much chemistry is required per 14 inches of film travel. As a general rule, the average processor will require 60–70 cubic centimeters (cc) of developer and 100–110 cc of fixer per 14 inches of film travel. The processor maintenance guide will state how the replenishing pumps are adjusted to increase or decrease replenishing rates. As a general rule, the more films that are processed per day, the less the amount of chemicals that need to be replaced; and vice versa. Overreplenishing or underreplenishing is not only wasteful but will affect sensitometry. Overreplenishing causes a slight increase in contrast, but there is little effect on the overall quality of the film. Underreplenishing, if the condition exists long enough, will lead to a reduction in the pH of the chemistry, with decreasing chemical activity as a result. The change causes an increase in the bromide level, and film speed and contrast may be greatly reduced.

Replenishment is achieved by the activation of a micro-switch as a film is fed into the processor through the feed tray; the switch remains activated until the film passes from the feed tray into the processor. To measure the replenishing rate, disconnect the replenishing line at its high point in the processor and catch the chemistry in a graduated cylinder while a 14-inch film is being fed. Adjustments should be made in accordance with the film manufacturer's specifications.

Replenishing tanks

Unfortunately, replenishing tanks are too often ignored in the radiology department. The following points should be considered.

1. The replenishing tanks should be outside the processor so that the chemicals will not be affected by the heat produced by the processor heater and dryer system. They should also be located within a convenient distance of the processor.

2. The tanks should be cleaned on a periodic basis. If chemical contamination is suspected, the remaining chemistry should be dumped and fresh chemistry installed.

3. The size of the tank should be sufficient to keep an adequate amount of fresh chemistry. Any chemistry remaining after approximately two weeks should be dumped because of aerial oxidation contamination.

4. Each replenishing tank should have a floating lid and dust cover to reduce loss of activity due to aerial oxidation.

5. Mixing paddles should be clearly marked, and under no circumstances should a paddle marked for one chemical system be used to mix the chemistry for the other system. No amount of cleaning will remove all traces of fixer (or developer) from a paddle; the use of a paddle from one system in another system will guarantee cross-contamination.

6. The height of the replenishing tank from the floor should be kept constant,

since head-pressure transmission from the tank will affect replenishing rates with certain types of replenishing pumps.

7. If more than one processor is replenished from a single tank, care should be taken that the replenishing pumps have sufficient reserve to compensate for the added resistance of longer replenishing lines.

Replenishing pumps

There are five types of replenishing pumps used in popular x-ray film processors.

Diaphragm-type replenishing pumps. This is a direct-metering pump which works on the principle of filling a flexible container to a certain volume and then forcing the fluid out. The diaphragm-type pump is greatly affected by head-pressure transmission from the replenishing tanks, and replenishing rates may vary as much as ±50 percent, depending on the head pressure. (Head pressure is the force exerted at the end of the replenishing line as the fluid comes from the replenishing tank. It is influenced by the height of the replenishing tank, the volume of fluid within the tank, the internal diameter of the replenishing line, and a number of other factors.) When a processor has this type of pump, the replenishing rate should be established when the replenishing tank is half-full. This rate causes overreplenishing 50 percent of the time (when the tank is full) and underreplenishing the rest of the time. However, the over- and underreplenishing usually averages out, so that film quality shows no significant change.

Bellows-type pump. This pump works on a principle similar to that of a fireplace bellows: when the bellows is opened, it fills with fluid and when the bellows closes, the fluid is expelled into the processor. This type of pump is affected only by a 5 percent error in head-pressure transmission. This is the recommended type of replacement pump; it is less expensive and more accurate than diaphragm-type pumps.

Magnetic-drive centrifugal pump. This pump is used mostly on the Kodak processors. It is a fixed-output pump, regulated by means of a separate needle valve. The pump is sensitive to head-pressure transmission, but less so than the diaphragm-type pump.

Percentage time pump. This type of pump is used on most Litton/Xonics processors. In this type, film feeding time is accumulated on a timing device, and every so often a high volume of chemistry is pumped into the processor to compensate for all the films fed into the processor during a set time period. This higher pumping rate and type of pump reduces the effect of head pressure.

Piston-type pumps. These are pumps of a new type which work on the principle of filling a syringe and then ejecting the fluid into the processor. The pumps are not affected by head-pressure transmission.

THE CIRCULATION/FILTRATION SYSTEM

Circulation pumps or agitating pumps provide for the movement of the chemistry through filters and for uniformity of the chemical activity throughout the tank, including uniformity of chemical components, temperature, and agitation. Filtration of the developer is required because accumulation of gelatin on the rollers in the developing tank will cause abrasions and pressure marks (artifacts) on the film. Filters are not generally used in fixer, because fixer neutralizes the developer and the debris produced will not harm the film. Fixer also hardens the gelatin, but it will not coat the rollers and produce film artifacts. Nor are filters required for the wash tank, because the wash tank is open-ended, with a large volume of water flow; debris and chemical residue are carried down the drain.

Filters

Water is supplied from the incoming water line in such a way that all particles larger than 90 microns are filtered out. The incoming water has to be filtered further for use in processing. For general purposes, a 40-micron filter can be used. A 30-micron filter is used to remove ion oxide particles. If the water supply is particularly dirty, filters should be added in tandem, with a large micron-size filter first, followed by a smaller-micron filter as the second one. In the case of dirty water, if a small filter only is used, it will clog easily and require early replacement, which is expensive. To remove the algae requires a filter of 1-micron size.

Circulation pumps

Proper agitation is required for good sensitometric results. Lack of agitation may create hot or cold spots in a tank or areas where replenisher chemistry can pool, resulting in non-uniform development because oxidized and fresh developer are not rapidly and efficiently exchanged within the gelatin layer. Common causes of inadequate agitation are failure of circulation pumps or motor, a kink in the replenishing lines, an air lock in the lines, blockage of intake or outlet of pumps by foreign material, or clogging of filters in the developer circulation system. Failure of adequate circulation in the developer system results in failure of the hydroquinone activity and thus in an undeveloped film. This appears as loss of contrast, decrease in maximum density, and loss of speed.

THE DRYER SYSTEM

Besides drying the radiographic film, the dryer system facilitates handling of the film. The system works in this fashion. A blower draws in room air which is then heated and circulated through a series of air tubes to the compartments of the dryer where heated air is directed against the wet film emerging from the wash tank. Part of the exhaust air is recirculated in the dryer, and the cooler moist air is pushed out of the compartment through the exhaust. The intake air is forced over a heating element that heats incoming air and increases its moisture capacity. Heating elements are generally of 2300- or 2500-watt rating and there are two or three in each processor. An adjustable thermostat regulates the temperature of the air by controlling the heat output of the heaters. A safety thermometer is included in most processors.

The dryer assembly does not ordinarily affect the sensitometric properties of the film, but it can affect the visualization of the image. If a film is improperly dried, air streaks—light and dark lines—appear on the film and will show up when the radiograph is viewed on the illuminator.

The quality of the room air that enters the dryer will affect the film. Debris from the air drawn through the dryer can cause artifacts; moist air can delay the drying. Dust can block the air flow and thus lower the efficiency of the dryer.

Two kinds of wet films are frequently encountered even when the dryer system appears to be functioning normally. One kind is due to chemical failure in the processing of the film, specifically a general lack of hardener. This can occur in the developer and the fixer through either chemical contamination or chemical depletion. Such films are soft and easily damaged. The second kind of wet film is due to failure of the dryer system to properly exhaust moisture. This occurs when a heater burns out, when the blower is not running at its normal speed

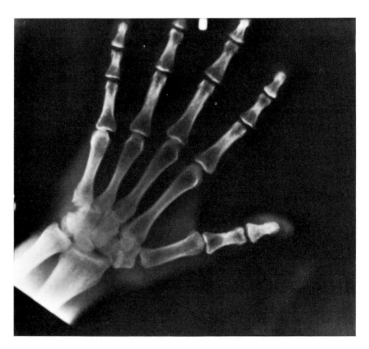

Figure 11-8. Note the pressure artifact lateral to the thumb on this exposure of a hand phantom.

(sometimes due to slipping drive belts), or when there is some obstruction in the ventilation system.

FILM ARTIFACTS

Artifacts can be classified as either sensitometric or physical defect on the film. They are unwanted, useless images that will interfere with radiologic interpretation. Listed below, in order of frequency, are the general sources of the artifacts:

1. Exposure
2. Processing
3. Handling
4. Storage
5. Manufacturing

Artifacts can also be classified as either sensitized or nonsensitized. Sensitized artifacts result from any type of exposure that sensitizes the emulsion—static electricity, pressure, extraneous radiation, and some others. Nonsensitized artifacts result from physical damage to the film—dirt deposits, scratches, and the like.

Manufacturing artifacts

Film manufacturers have rigid quality-control standards. Some companies stencil emulsion numbers on their film, and other identifiable marks, so as to be able to trace film and correct the manufacturing defects.

Storage artifacts

Some artifacts result from the way film is stored. Fresh film should be stored in an area where the temperature is 15–21° centigrade (60–70° Fahrenheit) and the relative humidity is 50–60 percent. Boxes containing film should be placed upright or on edge so as to provide maximum support and should be stacked no more than three high. Storing film flat may produce pressure artifacts (Figure 11-8). The film storage area should be free from extraneous radiation and heat and chemical fumes. The stock should be rotated, on the first in/first out principle (FIFO); i.e., the oldest should be used first. Records of the film used should be kept to maintain the system.

Handling artifacts

Improper handling of film during transportation can be a serious cause of artifact production. Increased fog, pressure artifacts, and sensitization are common problems associated with transportation handling. Perspiration from fingers, fingernail marks, and static discharges can occur with improper handling in the darkroom, prior to processing. Static discharges are considered handling artifacts. Three distinct types of sensitized artifacts caused by static electricity are seen: (1) tree static, which produces multiple branching dark lines: (2) ground static, which produces multiple branching dark lines emanating from a broad base, i.e., crownlike; and (3) smudge static, which produces an irregular dark blotch on the film. Static discharge can usually be controlled by having equipment properly grounded and maintaining an adequate relative humidity in the darkroom, possibly 50 percent.

Exposure artifacts

Exposure artifacts are generally related to improper exposure techniques. Other exposure artifacts include dirt or foreign materials in or on the screens, poor screen contact, warped or bent cassettes, light leaks around the cassettes, and improper combination of screens and film.

Processing artifacts

The most common processing artifacts arise from improper relationship between the time, temperature, and chemical activity factors. To identify processing artifacts, the following sequence is helpful.

A. Process a fresh, unexposed 14 × 17 in. film and mark the leading edge, side, date, time, front, and back of the film after processing. The film should emerge from the processor completely clear. If the film has chemical fog, it may be defective (i.e., preprocessing fog), or the chemical system may be at fault. If fog is present, the processor should be cleaned and adjusted, fresh chemistry installed, and the process repeated.

B. Process a 14 × 17 in. *white*-light exposed film (this should be flashed, since overexposure will cause a reversal effect). This film should emerge from the processor totally black. If the surface of the film as seen in reflected light is brown, this indicates developer oxidation. Usually if the chemistry is contaminated or old, this would show up in a finished radiograph as a decrease in contrast and a loss of film speed on the radiograph.

C. Process a 14 × 17 in. *x-ray* flash exposure film (70 kV, 2 mAs, tabletop). This film should emerge from the processor with a uniform light density. This radiograph duplicates the extrasensitivity the film has after exposure and thus is similar to actual radiographic exposures. The films described under A and B are not like films of patients.

D. To locate processing artifacts related to the transport mechanism, a marked film should be run through each assembly and inspected for scratches, gashing marks, pressure marks, etc. Each rack must then be visually inspected to correct the defect.

Pi lines. Pi lines, so called because they occur 3.14 times the diameter of a roller away from the leading edge of a film (see Figure 11-9), are most common in newly installed or freshly cleaned processors. They tend to disappear with use of the processor.

Guide shoe marks. Guide shoe marks—regularly spaced scratches or high-density lines (see Figure 11-10)—are due to improperly adjusted guides in a processor. Check clearances between guide devices and adjacent rollers or other components.

Dirt deposits. Irregular deposits, often light in color, generally elongated in the direction of film travel (see Figure 11-11), are caused by dirt or precipitate in water supplied to washing section. If the condition is temporary, clean the wash rack and replace the wash water in the processor; drain the wash tank when shutting the processor down. If the condition persists, change filters in

gure 11-9. "Pi line," thought to be caused by minute deposit left on a roller by the leading edge a film. The deposit is then carried around and nsferred to the film. (Courtesy, Radio Markets vision, Eastman Kodak Company.)

Figure 11-10. Guide marks are caused by improperly adjusted film guides in automatic processors. They are regularly spaced and in the direction of film travel. (Courtesy, Radio Markets Division, Eastman Kodak Company.)

Figure 11-11. Surface deposits can be caused by dirt in the wash water supplied to an automatic processor. (Courtesy, Radio Markets Division, Eastman Kodak Company.)

the incoming water lines. Some dirt deposits can be removed from the dry radiograph by gentle rubbing with a dry, soft cotton cloth.

SILVER RECOVERY UNITS

In times past it was common for the chief technologists in the x-ray department to be given the silver which could be recovered from the processing of the radiographic film. The amount of money recovered from selling reclaimed silver was of no great magnitude. However, nowadays the scarcity and high price of silver require that virtually all radiographic processing units have some type of silver-recovery unit attached to them. (The high price of the metal has also led many companies into research for alternatives or substitutes which can be used in lieu of silver halide in image formation.)

It is the function of the fixer to clear unexposed and undeveloped silver bromide crystals from the film emulsion, in addition to shrinking and hardening the emulsion and stopping the developing action. The fixer is replenished so as to maintain a fairly constant level of pH and of other fixer chemicals. As a result of the steady carry-in of traces of materials from the developing tank into the fixer the steady carry-out of materials from the fixer to the wash tank, there is a steady increase in the silver content of the fixer solution. Thus the fixer gradually weakens, both through the using-up of its ingredients and through dilution by carry-in and carry-out. In addition, as silver content of the fixer increases, the time required to fix or clear the film also increases.

The purpose of continuous silver recovery is to remove the silver, thereby keeping fixer time fairly constant. The chemical activity of the fixer is maintained by replenishing the fixer with fresh fixer solution.

There are two basic methods used in recovering silver from a processing unit: chemical and electrolytic.

Chemical recovery units

There are two chemical methods of silver recovery: (a) precipitation of the silver or (b) replacement of the silver by another metal.

(a) In the precipitation method, the silver in the fixer solution is precipitated (as silver sulfide) by adding sodium sulfide to the solution. The reaction between sodium sulfide and silver bromide creates silver sulfide, which precipitates out of solution. This method of recovery is highly efficient, but the efficiency is offset by the production of toxic fumes and of silver sulfide sludge, which is difficult to handle and dry. Chemical precipitation is rarely used in the radiographic department.

(b) The other chemical method of silver reclamation is with the use of some metal to replace the silver. If, for example, steel wool is placed in a tank of expended fixer, the iron ions of the steel wool will go into the solution and the silver in the solution will replace the iron ions from the steel wool. Thus, the silver in the solution will plate out on the steel wool.

This is an inefficient method and the presence of the iron (ferrous) ions will make the fixer unsuitable for re-use.

Thus the chemical methods of silver recovery, both (a) and (b) above, make the fixer unsuitable for re-use in the processing of x-ray film. In order to use chemical methods of reclaiming silver, the expended fixer is collected from the processor and stored in a container until the silver is recovered. After recovery of the silver, the solution is discarded.

Electrolytic recovery units

The most common method of silver recovery used in radiography is the electrolytic method, or electrolysis. Electrolysis is the process of producing a chemical change by passing an electrical current through an ionized solution. Whereas the chemical recovery units render a fixer unsuitable for re-use, the electrolytic methods allow the fixer to be re-used and still show a high efficiency in recovery of the silver (about 95 percent).

In electrolytic recovery units, the fixer is stored in a container outside the processor, and silver is recovered by passing a direct current through it. Two electrodes are used: the anode (positive pole) is usually made of carbon, and the silver ions move from it towards the cathode, which is normally made of stainless steel. Since the silver exists in the fixer as positively charged ions, it will be attracted to the cathode (the negative pole) and will be plated on the stainless steel there. Agitation of the fixer solution heightens not only the efficiency of removal of the silver but also the speed with which the silver is deposited on the cathode. Agitation of the solution also allows the use of higher direct-current voltages; non-agitation requires the use of lower voltages. Recovery units that employ low current—so-called low-current units—are relatively inexpensive and are for use in small processors; high-current (agitated) units are more expensive and are for use where large volumes of fixer are involved.

The decision as to recirculation of the fixer and silver removal depends entirely on the volume of fixer being used and the cost of the equipment and the labor required to recirculate the fixer. In small-volume radiographic areas, it is generally not economical to recirculate the fixer; after the silver is recovered, the fixer is discarded. In very large units, it is sometimes more economical to pump the fixer from which the silver has been recovered back to the fixer-replenishing holding tank for re-use. There is much disagreement about whether to recirculate fixer. When fixer from which the silver has been removed is recirculated, the silver is gone, but not the bromide, and it is the bromide, not the silver, which retards fixation.

COMMON FILM AND PROCESSING PROBLEMS

Tables 11-3 and 11-4 indicate some of the common problems seen in a radiology department. Although the tables indicate where to look for some of these problems, the student should understand that *all variables are to be checked before any adjustments are made*. It is common, for example, for someone to increase the temperature of the developer to increase the speed (density) of the film; but the problem in an underprocessed film may have been a contaminated developer, such as may occur, for example, when the fixer rack is moved and a few cc's of fixer inadvertently fall into the developer tank. If all factors have been working and there is an abrupt change, look for simple remedies rather than complex ones. When a film appears underexposed, the error is much more likely to be that the operator has picked up a cassette with a medium-speed screen, when he thought it had a high-speed screen, than that the exposure timer has suddenly gone out of calibration. Most of the difficulties indicated in the tables can be corrected with an insight into potential common problems and attention to the details of the examination and the processing of the film.

Experiment 1. Effect of temperature of developer on density.

Purpose. To demonstrate the fact that as temperature increases, development time must be decreased to compensate for increased activity.

Procedure. Expose 4 postero-anterior radiographs of the chest phantom with the following factors:

Film:	Any 35 × 43 cm (14 × 17 in.)
Screens:	Medium-speed
Grid:	None
Distance:	183-cm SID (72 in.)
Collimate:	To phantom
kV:	80
mAs:	10
Development:	Variable factor

Variable factors:

Exposure 1:	Develop 8½ min at 60°
Exposure 2:	Develop 5 min at 68°
Exposure 3:	Develop 3½ min at 75°
Exposure 4:	Develop 2 min at 80°
	Manual processing

Experiment 2. Effect of development and density.

Purpose. To demonstrate that as development time increases, density increases.

Procedure. A series of 10 radiographs of the postero-anterior chest phantom are made with the following factors:

Film:	Any 35 × 43 cm (14 × 17 in.)
Screen:	Medium-speed
Grid:	None
Distance:	183-cm SID (72 in.)
Collimate:	To phantom
kV:	80
mAs:	10
Development:	Variable factor

Variable factors:

Exposure 1:	Develop 1 min at 68°
Exposure 2:	Develop 2 min at 68°
Exposure 3:	Develop 3 min at 68°
Exposure 4:	Develop 4 min at 68°
Exposure 5:	Develop 5 min at 68°
Exposure 6:	Develop 6 min at 68°
Exposure 7:	Develop 7 min at 68°
Exposure 8:	Develop 8 min at 68°
Exposure 9:	Develop 9 min at 68°
Exposure 10:	Develop 10 min at 68°
	Manual processing

Table 11-3. Common problems with mechanical processors

Problem	Cause	Corrective Action
1. Decreased density	Underreplenishment	Check replenishing rates. Check for air locks in replenishing lines. Check for clogged filters.
	Developer temperature low	Check thermometer, thermostat. Check incoming water volume and temperature. Check heat exchanger. Check mixing valve.
	Decreased developer activity	Check for contaminated or exhausted developer. Check whether chemistry is mixed properly.
2. Increased density	Overreplenishment	Check replenishing rates.
	Developer temperature high	Check thermometer, thermostat. Check incoming water volume, pressure, and temperature.
	Overactive developer	Check mixing procedure and sequence. Check for not enough starter. Check whether replenishing rates are too high. Check for wrong chemistry.
	Immersion time prolonged	Check transport system. Look for excessive wear of gears and sprockets, alignment, lack of lubrication
3. Films jamming	Chemical	Check for improperly mixed chemicals. Check for contaminated chemicals, especially fixer. Check for imbalance between fixer and developer temperature. Check for exhausted fixer.
	Mechanical	Check for malalignment of processing racks and crossovers. Check for dirty racks. Check for drying temperature too low. Check incoming dryer air—too moist or too cool. Check for worn drive gears or sprockets, causing hesitation. Check for loose roller retention springs.
4. Artifacts	Mechanical	Check for malaligned or dirty guide shoes. Check for dryer tubes not seated properly. Check for dirty rollers. Check for dirty wash water.

Table 11.3. cont.

Problem	Cause	Corrective Action
5. Films not drying	Mechanical	Check for dryer tubes not seated correctly. Check for incorrect drying temperature. Check dryer thermostat. Check for high humidity or low temperature of incoming air.
	Chemical	Check for lack of proper fixation, particularly lack of hardener. Check whether processing temperature is too high for the chemicals being used.

Table 11-4. Common film problems

Problem	Cause	Correction
Too little density	Underexposure	Wrong exposure factors.[a] Recheck. Check exposure timer.
	Grid	Check tube centering, focal length of grid, grid ratio.
	Equipment	Valve or timer failure. Line voltage too low. Gassy x-ray tube. Added beam filtration.
	Wrong film/screen combination	Recheck combination and adjust exposure factors.
	Underprocessing	Check Table 11–4
Too much density	Overexposure	Wrong exposure factors. Recheck.
	Fog	Check lack of grid, outdated film, overprocessing.
	Line surge	Recheck line compensator.
	Overprocessing	Check Table 11–4.
No exposure	No x-ray	Check: Fuses; Circuit breakers; voltage selector for dead buttons; Interlock switches; Under-or over-table tube selector; Tube overload circuits.

[a] Primary factors: kilovoltage; milliamperage; time; source/image-receptor distance.

SUGGESTED READING

American National Standards Institute Standard Ph2.9: "Method for the Sensitometry of Medical X-ray Films and Dental X-ray Films," 1974.

Burnett, B. M., Mazaferro, R. J., and Church, W. W. *A Study of Retakes in Radiology Departments of Two Large Hospitals.* USDHEW/PHS Publication No. (FDA) 76–8016.

Cope, D. J.: "Automatic Processing. A method of economising on Essential Services." *Radiography* 36 (1970):13–14.

Dobrin, R., Kricheff, I., Fite, W., and Weathers, R. "The Effect of Variability of Automatic Film Processing Systems on the Quality of Radiographs." *Radiology* 113 (1974):545–54.

Faix. C. D., Van Tuinen, R. J., and Kereiakes, J. G. "Quality control for automated film processing." *Radiol. Technol.* 44 (1973):257–61.

Ferguson, J. P., and Schadt, W. W. *Variability in the Automatic Processing of Medical X-ray Film.* USPHS/DHEW Publication No. BRH/DEP 70–13, 1970.

Gaynor, L. L., et al. "Who Needs Darkrooms?" *Radiol. Technol.* 47 (1976):237–44.

Gray, J. E. *Photographic Quality Assurance in Diagnostic Radiology, Nuclear Medicine, and Radiation Therapy,* Vol. 1. DHEW Publication No. (FDA) 76–8043, 1976.

Huff, K., and Shaffer, E. *Report of ANSI PH2.31 task force on sensitometry of screen-film-processing combinations.* Proceedings of Symposium, Medical X-ray Photo-Optical Systems Evaluation, DHEW Publication No. (FDA) 76–8020, 1974.

Iannazzi, R. F. "Processor Quality Control." *Photo Methods for Industry,* April 1973, 29–31.

Kelley, J. P., et al. A system of Periodic Testing of Film Processing. *Radiol. Technol.* 43 (1971):15–19.

Lawrence, D. J. "A Simple Method for Processor Control." *Med. Radiog. and Phot.* 49 (1973):2–6.

Lorimer, D. "Practical Aspects of QC of Processing X-ray and Cine Film." *Radiography* 40 (1974):250–63.

Mazzaferro, R. J., Balter, S., and Janower, M. L. The Incidence and Causes of Repeated Radiographic Examinations in a Community Hospital." *Radiology* 112 (1974):71–72.

Nichols, R. L., and Moseley, R. D. "Simple Sensitometric Control of Developer Activity." *Amer. Jour. Roentgenol.* 78 (1957):145.

Paznauski, A. K., and Smith, L. A. "Practical Problems in Processing Control." *Radiology* 90 (1968):135–38.

Thompson, T. T., Kirby, C. C., and McKinney, W. E. J. *A Guide for Automatic Processing and Film Quality Control.* The American Society of Radiologic Technologists, Chicago, 1975.

Van Tuinen, R. J., and Kereiakes, J. G. *Sensitometric Quality Control for Automated Film Processors in Radiology Departments.* Proceedings of Symposium, Reduction Dose in Diagnostic X-ray Procedures, DHEW Publication No. (FDA) 73–8009, 1972.

Winkler, N. T. *Quality Control in Diagnostic Radiology.* Proceedings of Symposium, Application of Optical Instrumentation in Medicine, IV, SPIE/SPSE, 1975.

REVIEW QUESTIONS

1. The average developer immersion time in an automatic "90-second" processor is approximately

    a. 15 sec   d. 30 sec
    b. 20 sec   e. 35 sec
    c. 25 sec

2. For orthochromatic-sensitive x-ray film, a suitable darkroom light is

   a. Wratten 6B
   b. Kodak GBX
   c. GE daylight blue
   d. Sylvania huelight
   e. Kodak OX

3. Film density in development is primarily related to

   a. Time of development
   b. Temperature of development
   c. Activity of developer
   d. A, B, and C are correct
   e. Only A and B are correct

4. The average replenishing rate for fixer is approximately

   a. 50–60 cc/film
   b. 100–110 cc/film
   c. 25–30 cc/film
   d. 200–220 cc/film

5. One of the most common causes of wet film emerging from an automatic processor is

   a. General lack of hardener in developer or fixer
   b. Wash water dirty
   c. Developer temperature too low
   d. Dryer temperature too high
   e. Processor drain clogged

6. The most common type of film artifact is

   a. Manufacturing error
   b. Exposure error
   c. Handling error
   d. Storage error
   e. Processing error

7. Fresh x-ray film should be stored at approximately

   a. 50° F/80% relative humidity
   b. 60–70° F/25% relative humidity
   c. 20–30° F/20–25% relative humidity
   d. 60–70° F/50–60% relative humidity
   e. 70–80° F/80–90% relative humidity

8. The most common method of silver recovery is the

   a. Electrolytic method
   b. Chemical method
   c. Steel wool method
   d. Precipitation method
   e. Osmotic method

9. Starter solution is composed of

   a. Potassium bromide   d. Potassium bromate
   b. Silver nitrate   e. Alum
   c. Silver sulfide

10. Starter solution is used to

    a. Increase fixer activity
    b. Increase developer activity
    c. Decrease fixer activity
    d. Decrease developer activity
    e. Stabilize wash water

11. Using a water hose to mix processing solutions

    a. Causes chemicals to splatter from one tank to another
    b. Contaminates the water hose
    c. Introduces air bubbles which cause oxidation
    d. Increases the temperature of the developer
    e. Has no effect on the developer

12. The most common reason for contaminated developer is

    a. Mixing fixer in the developer tank
    b. Processing at too high a temperature
    c. Mixing chemicals with a water hose
    d. Using a fixer rack in a developer tank
    e. Dripping fixer from the fixer rack into the developer tank when removing or replacing the fixer rack

13. Wash water is ordinarily kept at what temperature with respect to developer temperature?

    a. 5 F° above developer temperature
    b. 5 F° above fixer temperature
    c. 5 F° below developer temperature
    d. 15 F° above fixer temperature
    e. 15 F° below developer temperature

14. In an automatic processor, algae can be controlled by

    a. Adding 6 ounces of bleach to the wash water
    b. Adding starter solution to the fixer
    c. Adding 6 ounces of bleach to the developer
    d. Adding starter solution to the wash water
    e. Adding alum to the wash water

15. The normal fixing time in an automatic processor is

    a. The same as developing time
    b. The same as dry-to-drop time
    c. About twice developing time
    d. One half of washing time
    e. Double the film transport time

16. Dry-to-drop time is

    a. The time required for a dry film to drop three feet
    b. The time required for a film to drop from a dryer rack into the film bin
    c. The time it takes for a film to be fed through a processor and drop into the film receiving bin
    d. The time a film is in the developer solution
    e. The time a film is in the fixer solution

17. The function of developer is to

    a. Swell the emulsion
    b. Reduce exposed silver halide crystals

   c. Fix the latent image
   d. A, B, and C are correct
   e. Only A and B are correct

18. The developer used in an automatic processor is termed what type of developer?

   a. PQ   d. PR
   b. MQ   e. QR
   c. PM

19. What part of the developer chemistry is most affected by changes in temperature?

   a. Hydroquinone
   b. Phenidone
   c. Potassium bromide
   d. Sodium carbonate
   e. Sodium sulfite

20. The average replenishing rate for developer solution is

   a. 60–70 cc per film
   b. 100–110 cc per film
   c. 25–30 cc per film
   d. 200–210 cc per film

21. A crossover rack is located

   a. At the top of the vertical rack assembly
   b. At the bottom of the vertical rack assembly
   c. At the bottom of the dryer rack
   d. At the entrance of the film into the processor
   e. Not located in the processor

22. How often should a crossover rack be cleaned?

   a. As needed
   b. Daily
   c. Weekly
   d. Monthly
   e. Every time the processor is turned off

23. Guideshoes are

   a. Turnaround racks
   b. Planetary rollers
   c. Solar rollers
   d. Master rollers
   e. Deflector plates

24. Processing solutions should be dumped

   a. Whenever they are contaminated
   b. When oxidation occurs
   c. When crossover racks are cleaned
   d. Daily
   e. Weekly

25. Development temperature should be checked

   a. Periodically throughout the working period
   b. Once daily
   c. Weekly
   d. Monthly
   e. When something goes wrong

APPENDIX A. BIT SYSTEM OF TECHNIQUE CONVERSION

WHAT THE BIT SYSTEM DOES

The Bit System helps to change techniques, create new exposure and processing conditions, estimate patient roentgen dosage, and understand the factors affecting radiographic quality. It supplements practical experience by helping you make changes quickly and accurately.

Film-screen speed

The Bit values in Table P summarize the relative speeds of all Cronex® Film-Screen combinations.

How it works

Bit values are relative exposure units common to all radiographic variables and assigned according to the effect of the variable on film density. They serve the same trading purpose as money. Values are arranged in tables so that reading up the column increases film exposure and density and vice versa. A + 1.0 Bit increase doubles exposure and a −1.0 decrease halves exposure (see Table C). The following pages contain four ways to use the Bit System.

With Du Pont Bit System you can:

make changes with the speed and mathematical accuracy of simple addition and subtraction.

use either centimeter or inch distances; Fahrenheit or Centigrade temperatures.

be compatible with existing technique change methods. Table C lists equivalent % and mAs factor changes.

incorporate new radiographic variables as their control becomes desirable.

Bit tables are accurate under average conditions of use. However, normal differences in X-ray beam radiation quality, milliampere output, and other variables introduce errors into the Bit System as well as all other technique conversion systems. Usually, an error of 0.3 Bits (22%) can be tolerated in a radiograph.

FOUR WAYS TO USE THE BIT SYSTEM

1. Changing techniques by measuring on the tables

Technique charts frequently need to be changed due to individual preference,

Reprinted with permission of E. I. du Pont de Nemours & Co. (Inc.).

new equipment, film, screens, processing conditions or whatever. Some of the typical changes that can be made with the Bit System are:

Raise kVp and shorten time to stop motion.

Change from no grid to grid and raise kVp to increase exposure latitude.

Use faster screens and shorter time to reduce dosage.

Increase illuminator brightness and mAs to increase radiographic contrast.

Increase distance and kVp to reduce penumbra unsharpness.

Increase screen speed and reduce focal spot size to reduce penumbra unsharpness in magnification techniques.

Lower developer temperature and increase mAs to reduce quantum mottle.

Any of the above changes can be made by the following procedure. All radiographic variables are spaced on this card so that a distance change on one table will have the same effect on exposure as the same distance on another table.

Example: kVp increase and time decrease to reduce contrast

(1) Place edge of a piece of paper beside kVp Table H.
(2) Mark the paper edge at 64 and 85 kVp in this case.
(3) To indicate a kVp increase, place an up arrow between the marks on the paper.
(4) Move paper edge to "time" Table K and place the mark at the arrowhead end; beside the original (2/5 sec.).
(5) Read new exposure time (1/10 sec.) at the other mark.

2. Changing techniques by totaling Bits

Theoretically, it should not make any difference to the film what combination of technical factors within reason are selected to make an exposure as long as a certain total number of exposure units (Bits) are provided for a given density. Phototiming works in this manner.

Assuming you want the same film density as the original, total the Bits of the original factors to be changed and subtract the Bit Value of the new factor to be used. The remainder will be the Bit Value of the other new factor.

Example: kVp increase and time decrease to reduce contrast

Original kVp	:	64 =	6.4 Bits
Original sec.	:	2/5 =	+11.0
Original Total	:		17.4
New kVp	:	85 =	− 8.4
Unknown sec.	:		9.0 Bits or 1/10 sec.

3. Creating new techniques using Bit Totals provided

This extension of the totalling method discussed on the opposite page is used when techniques are nonexistent. A technique is selected to exactly meet the Guide Bit Total provided. Two applications of this principle follow:

Correct exposure exists when kVp Bits + mAs Bits + Film-Screen Bits + Grid Bits + F-F Distance Bits + Anatomical Thickness Bits = Guide Bit Total. These typical Guide Totals may be used to fomulate starting techniques:

Part and View		*Guide Total*
Lumbar Spine, Lateral	=	63.6 Bits
Abdomen	=	63.1
Skull, Lateral	=	62.3
Chest, Anterior	=	60.9
Hand, Anterior Oblique	=	51.7
(subtract 1.0 Bits from total for 3 phase generators)		

Your own Guide Totals may be obtained by totalling Bits on all important variables used to expose excellent radiographs. These Totals are then used when new techniques are needed.

Optimum development exists when Developing Time Bits + Temperature Bits + Developer Activity Bits = 15.0

Example: Obtaining optimum development

First, measure immersion time without crossovers. It is the time from leading edge under to leading edge out of the developer. Knowing immersion time and types of developer, proceed as follows using Table Q:

Optimum Guide Total	=	15.0 Bits
Known Activity: XMD	=	5.0
Known Immersion: 18 sec.	=	+4.0
Known Total	=	−9.0 Bits
Unknown Temperature		6.0 Bits
		or 92°F

Developing conditions totalling 15.0 Bits in Cronex® developers produce film speed, fog, contrast and background density that meet product design specifications. Above 15.0 Bits, quantum noise becomes worse. Below 15.0, patient dose is increased.

4. Estimating skin dose by totalling Bits

Skin dose is defined as that quantity of roentgens incident to skin surface. Depending on the energy wave-length distribution of the beam and the absorption characteristics of the tissue, only a fraction of the skin dose is actually absorbed. The remainder exposes the film. At low kVp, percent absorption is greatest. Because of the complexity of radiation absorbed dose (rad) determination, only skin dose is approximated here.

KVp, mAs and distance (Tables B, I and F) are used to estimate skin dose (Table A). The special kVp calibration applies with 3mm aluminum total filtration or equivalent and single phase, full wave rectification. For 3 phase or constant potential rectification, add .8 to the single phase Bit total. Follow this procedure:

Cahoon's Formulating X-Ray Techniques

(1) Obtain Focal-Skin Distance by subtracting patient thickness (cm) from the Focal-Film distance (cm).

(2) Add Focal-Skin Distance Bits, mAs Bits and kVp Bits.

(3) Read mr skin dose next to the above total located in the mr Bit table.

Example: 20 cm chest skin dose (single phase)

Focal-Film Distance = 72″	= 183 cm	
Chest Thickness	= −20 cm	
Focal-Skin Distance	= 163 cm =	5.4 Bits
	+ 10 mAs =	16.0
	+ 70 kVp =	7.8 *
	Total =	29.2 Bits = 23 mr

* Use Roentgen Estimating kVp Table B only.

TABLE A APPROXIMATE SKIN DOSE		TABLE B KVP = BITS	

Roentgens, (r)
Milliroentgens, (mr) Bits

For Estimating Roentgens Only

Roentgens/Milliroentgens	Bits	KVP	Bits
10.5 r	= 38.0	155	= 10.2
9.0	= 37.8	140	= 10.0
7.8	= 37.6	130	= 9.8
6.8	= 37.4	120	= 9.6
5.9	= 37.2	111	= 9.4
5.2	= 37.0	104	= 9.2
4.5	= 36.8	97	= 9.0
3.9	= 36.6	91	= 8.8
3.4	= 36.4	86	= 8.6
3.0	= 36.2	82	= 8.4
2.6	= 36.0	77	= 8.2
2.3	= 35.8	73	= 8.0
2.0	= 35.6	70	= 7.8
1.7	= 35.4	66	= 7.6
1.5	= 35.2	63	= 7.4
1.3	= 35.0	61	= 7.2
1.1	= 34.8	57	= 7.0
960 mr	= 34.6	55	= 6.8
840	= 34.4	53	= 6.6
730	= 34.2	50	= 6.4
640	= 34.0	48	= 6.2
560	= 33.8	46	= 6.0
485	= 33.6	44	= 5.8
420	= 33.4	42	= 5.6
365	= 33.2	41	= 5.4
320	= 33.0	39	= 5.2
280	= 32.8	38	= 5.0
240	= 32.6	36	= 4.8
210	= 32.4	34	= 4.6
182	= 32.2	33	= 4.4
160	= 32.0	32	= 4.2
140	= 31.8	31	= 4.0
120	= 31.6	29	= 3.8
105	= 31.4	28	= 3.6
92	= 31.2		
80	= 31.0		
70	= 30.8		
60	= 30.6		
52	= 30.4		
45	= 30.2		
40	= 30.0		
35	= 29.8		
30	= 29.6		
26	= 29.4		
23	= 29.2		
20	= 29.0		
18	= 28.8		
15	= 28.6		
13	= 28.4		
11	= 28.2		
10	= 28.0		
9	= 27.8		

3mm Aluminum or Equivalent
Total Filtration
Single phase—full wave rectified
Tungsten Tube

Cahoon's Formulating X-Ray Techniques

TABLE C

Bit Change	% Increase (+) % Decrease (−)	Dosage or mAs Factor*	Relative mAs or Arithmetic Speed
+5.0	+3100%	32.00 X	3200
+4.8	+2720%	28.20 X	2820
+4.6	+2360%	24.60 X	2460
+4.4	+2040%	21.40 X	2140
+4.2	+1760%	18.60 X	1860
+4.0	+1500%	16.00 X	1600
+3.8	+1310%	14.10 X	1410
+3.6	+1130%	12.30 X	1230
+3.4	+ 970%	10.70 X	1070
+3.2	+ 812%	9.12 X	912
+3.0	+ 700%	8.00 X	800
+2.8	+ 592%	6.92 X	692
+2.6	+ 503%	6.03 X	603
+2.4	+ 425%	5.25 X	525
+2.2	+ 357%	4.57 X	457
+2.0	+ 300%	4.00 X	400
+1.8	+ 247%	3.47 X	347
+1.6	+ 200%	3.00 X	300
+1.4	+ 163%	2.63 X	263
+1.2	+ 130%	2.30 X	230
+1.0	+ 100%	2.00 X	200
+ .8	+ 74%	1.74 X	174
+ .6	+ 51%	1.51 X	151
+ .4	+ 32%	1.32 X	132
+ .2	+ 15%	1.15 X	115
0	0	1.00 X	100
− .2	− 13%	.87 X	87
− .4	− 24%	.76 X	76
− .6	− 34%	.66 X	66
− .8	− 42%	.58 X	58
−1.0	− 50%	.50 X	50
−1.2	− 56%	.44 X	44
−1.4	− 62%	.38 X	38
−1.6	− 67%	.33 X	33
−1.8	− 71%	.29 X	29
−2.0	− 75%	.25 X	25
−2.2	− 78%	.22 X	22
−2.4	− 81%	.19 X	19
−2.6	− 83%	.17 X	17
−2.8	− 86%	.14 X	14
−3.0	− 87%	.13 X	13
−3.2	− 89%	.11 X	11
−3.4	− 91%	.09 X	9
−3.6	− 92%	.08 X	8
−3.8	− 93%	.07 X	7
−4.0	− 94%	.06 X	6
−4.2	− 95%	.05 X	5
−4.4	− 95.5	.045	4.5
−4.6	− 96%	.04 X	4
−4.8	− 96.5	.035	3.5
−5.0	− 97%	.03 X	3

*New mAs = Old mAs X mAs Factor

TABLE D
CRONEX* 4 FILM DENSITY AND CONTRAST (Screen Exposure)

Bits		Density		Contrast
19.0	=	3.27	=	
18.8	=	3.25	=	.30
18.6	=	3.23	=	.33
18.4	=	3.21	=	.36
18.2	=	3.18	=	.39
18.0	=	3.16	=	.42
17.8	=	3.13	=	.50
17.6	=	3.10	=	.58
17.4	=	3.06	=	.67
17.2	=	3.02	=	.83
17.0	=	2.96	=	1.00
16.8	=	2.90	=	1.33
16.6	=	2.80	=	1.67
16.4	=	2.70	=	1.92
16.2	=	2.57	=	2.50
16.0	=	2.40	=	3.08
15.8	=	2.20	=	3.33
15.6	=	2.00	=	3.67
15.4	=	1.76	=	3.85
15.2	=	1.54	=	3.83
15.0	=	1.30	=	3.50
14.8	=	1.12	=	3.33
14.6	=	.90	=	3.33
14.4	=	.72	=	2.50
14.2	=	.60	=	1.92
14.0	=	.49	=	1.67
13.8	=	.40	=	1.17
13.6	=	.35	=	.75
13.4	=	.31	=	.58
13.2	=	.28	=	.42
13.0	=	.26	=	.33
12.8	=	.24	=	.25
12.6	=	.23	=	.17
12.4	=	.22	=	.08
12.2	=	.21	=	.01
12.0	=	.20	=	

*An average gradient over a .4 Bit exposure range.

TABLE E
X-RAY GENERATOR RECTIFICATION (usually change kVp)

Three-phase, 12 pulse = 5.0

Three-phase, 6 pulse = 4.8 / 4.6

= 4.4

= 4.2

Single-phase, full wave = 4.0

TABLE F
FOCAL-FILM DISTANCE

Inches	=	Bits	=	Centimeters
6.7	=	12.0	=	17
7.3	=	11.8	=	18.5
7.7	=	11.6	=	19.5
8.3	=	11.4	=	21
9.0	=	11.2	=	22.5
9.5	=	11.0	=	24
10.0	=	10.8	=	25.5
10.6	=	10.6	=	27
11.4	=	10.4	=	29
12.2	=	10.2	=	31
13	=	10.0	=	33
14	=	9.8	=	35
15	=	9.6	=	38
16	=	9.4	=	40
17	=	9.2	=	43
18	=	9.0	=	46
19	=	8.8	=	48
21	=	8.6	=	53
22	=	8.4	=	56
24	=	8.2	=	61
25	=	8.0	=	64
27	=	7.8	=	69
29	=	7.6	=	74
31	=	7.4	=	79
33	=	7.2	=	84
36	=	7.0	=	91
39	=	6.8	=	99
41	=	6.6	=	104
44	=	6.4	=	112
48	=	6.2	=	122
51	=	6.0	=	130
55	=	5.8	=	140
58	=	5.6	=	147
63	=	5.4	=	160
67	=	5.2	=	170
72	=	5.0	=	183
77	=	4.8	=	196
83	=	4.6	=	211
88	=	4.4	=	224
95	=	4.2	=	241
102	=	4.0	≈	259
109	=	3.8	=	277
116	=	3.6	=	295
125	=	3.4	=	318
136	=	3.2	=	345
144	=	3.0	=	366
154	=	2.8	=	391
165	=	2.6	=	419
176	=	2.4	=	447
180	=	2.2	=	457
203	=	2.0	=	516
218	=	1.8	=	554

TABLE G
ANATOMICAL THICKNESS

Centimeters	=	Bits
1	=	10.8
2	=	10.6
3	=	10.4
4	=	10.2
5	=	10.0
6	=	9.8
7	=	9.6
8	=	9.4
9	=	9.2
10	=	9.0
11	=	8.8
12	=	8.6
13	=	8.4
14	=	8.2
15	=	8.0
16	=	7.8
17	=	7.6
18	=	7.4
19	=	7.2
20	=	7.0
21	=	6.8
22	=	6.6
23	=	6.4
24	=	6.2
25	=	6.0
26	=	5.8
27	=	5.6
28	=	5.4
29	=	5.2
30	=	5.0
31	=	4.8
32	=	4.6
33	=	4.4
34	=	4.2
35	=	4.0
36	=	3.8
37	=	3.6
38	=	3.4
39	=	3.2
40	=	3.0
41	=	2.8
42	=	2.6
43	=	2.4
44	=	2.2

TABLE H

kVp	=	Bits
153	=	11.4
147	=	11.2
141	=	11.0
135	=	10.8
130	=	10.6
125	=	10.4
120	=	10.2
115	=	10.0
110	=	9.8
106	=	9.6
102	=	9.4
98	=	9.2
94	=	9.0
91	=	8.8
88	=	8.6
85	=	8.4
82	=	8.2
80	=	8.0
78	=	7.8
76	=	7.6
74	=	7.4
72	=	7.2
70	=	7.0
68	=	6.8
66	=	6.6
64	=	6.4
62	=	6.2
60	=	6.0
58	=	5.8
56	=	5.6
55	=	5.4
54	=	5.2
53	=	5.0
51	=	4.8
50	=	4.6
49	=	4.4
47	=	4.2
46	=	4.0
45	=	3.8
44	=	3.6
43	=	3.4
42	=	3.2
41	=	3.0
39	=	2.8
38	=	2.6
37	=	2.4
35	=	2.2
34	=	2.0
33	=	1.8
31	=	1.6
30	=	1.4
29	=	1.2
28	=	1.0
27	=	.8
26	=	.6
25	=	.4

TABLE I

mAs	=	Bits
640	=	22.0
550	=	21.8
485	=	21.6
420	=	21.4
365	=	21.2
320	=	21.0
280	=	20.8
240	=	20.6
210	=	20.4
182	=	20.2
160	=	20.0
140	=	19.8
120	=	19.6
105	=	19.4
92	=	19.2
80	=	19.0
70	=	18.8
60	=	18.6
52	=	18.4
45	=	18.2
40	=	18.0
35	=	17.8
30	=	17.6
26	=	17.4
23	=	17.2
20	=	17.0
17.5	=	16.8
15	=	16.6
13	=	16.4
11.2	=	16.2
10	=	16.0
8.7	=	15.8
7.5	=	15.6
6.6	=	15.4
5.7	=	15.2
5.0	=	15.0
4.4	=	14.8
3.75	=	14.6
3.33	=	14.4
2.90	=	14.2
2.50	=	14.0
2.16	=	13.8
1.80	=	13.6
1.66	=	13.4
1.45	=	13.2
1.25	=	13.0
1.08	=	12.8
.94	=	12.6
.83	=	12.4
.72	=	12.2
.63	=	12.0

Cahoon's Formulating X-Ray Techniques

TABLE J

mA	=	Bits
2100	=	11.4
1840	=	11.2
1600	=	11.0
1400	=	10.8
1200	=	10.6
1060	=	10.4
900	=	10.2
800	=	10.0
700	=	9.8
600	=	9.6
530	=	9.4
450	=	9.2
400	=	9.0
350	=	8.8
300	=	8.6
260	=	8.4
225	=	8.2
200	=	8.0
175	=	7.8
150	=	7.6
130	=	7.4
112	=	7.2
100	=	7.0
87	=	6.8
75	=	6.6
65	=	6.4
56	=	6.2
50	=	6.0
43	=	5.8
38	=	5.6
33	=	5.4
28	=	5.2
25	=	5.0
22	=	4.8
19	=	4.6
16	=	4.4
14	=	4.2
12	=	4.0
11	=	3.8
9.5	=	3.6
8.3	=	3.4
7.2	=	3.2
6.3	=	3.0
5.4	=	2.8
4.7	=	2.6
4.1	=	2.4
3.6	=	2.2
3.2	=	2.0
2.7	=	1.8
2.35	=	1.6
2.05	=	1.4
1.80	=	1.2
1.55	=	1.0
1.35	=	.8
1.17	=	.6
1.02	=	.4

TABLE K

Fraction Seconds	=	Bits	=	Decimal Seconds
6	=	15.0	=	6.32
		14.8	=	5.50
5	=	14.6	=	5.00
4	=	14.4	=	4.20
		14.2	=	3.70
3	=	14.0	=	3.20
		13.8	=	2.80
2-1/2	=	13.6	=	2.50
2	=	13.4	=	2.10
		13.2	=	1.83
1-1/2	=	13.0	=	1.60
		12.8	=	1.40
1-1/5	=	12.6	=	1.20
1	=	12.4	=	1.05
		12.2	=	.920
3/4	=	12.0	=	.800
		11.8	=	.700
3/5	=	11.6	=	.600
1/2	=	11.4	=	.525
		11.2	=	.455
2/5	=	11.0	=	.400
3/10	=	10.8	=	.350
		10.6	=	.300
1/4	=	10.4	=	.265
		10.2	=	.230
1/5	=	10.0	=	.200
		9.8	=	.175
3/20	=	9.6	=	.151
2/15	=	9.4	=	.132
		9.2	=	.115
1/10	=	9.0	=	.100
1/12	=	8.8	=	.087
		8.6	=	.075
1/15	=	8.4	=	.066
		8.2	=	.057
1/20	=	8.0	=	.050
1/24	=	7.8	=	.044
		7.6	=	.038
1/30	=	7.4	=	.033
		7.2	=	.029
1/40	=	7.0	=	.025
1/50	=	6.8	=	.022
		6.6	=	.019
1/60	=	6.4	=	.0166
1/72	=	6.2	=	.0143
1/80	=	6.0	=	.0125
1/90	=	5.8	=	.0110
1/105	=	5.6	=	.0095
1/120	=	5.4	=	.0083
1/140	=	5.2	=	.0072
1/160	=	5.0	=	.0063
1/180	=	4.8	=	.0053
1/215	=	4.6	=	.0046
1/250	=	4.4	=	.0040
1/280	=	4.2	=	.0035
1/333	=	4.0	=	.0030
1/360	=	3.8	=	.0026
1/430	=	3.6	=	.0023
1/500	=	3.4	=	.0020
1/560	=	3.2	=	.0017
1/666	=	3.0	=	.0015
1/780	=	2.8	=	.0013
1/860	=	2.6	=	.0012
1/1000	=	2.4	=	.0010

TABLE L
CHANGES REQUIRED FOR PATHOLOGY

RESPIRATORY SYSTEM

	Bits
Atelectasis	+.4
Bronchiectasis	+.4
Carcinoma (advanced)	+.4
Edema	+.4
Empyema	+.6
Hydropneumothorax	+.6
Pleural Effusion	+.6
Pneumoconiosis	+.5
Pneumonia	+.5
Thoracoplasty	+.6
Tuberculosis (calcific-miliary)	+.4
Emaciation	−.5
Emphysema	−.6

OTHER SYSTEMS

Acromegaly	+.4
Ascites	+.6
Arthritis (Rheumatoid)	+.4
Cirrhosis of Liver	+.4
Charcot Joint	+.4
Edema	+.4
Hydrocephalus (without air)	+.4
Metastasis	+.4
(Secondary to prostatic carcinoma)	
Osteochondroma	+.4
Osteoma	+.4
Osteomyelitis (healed)	+.4
Osteopetrosis	+.4
Paget's Disease	+.4
Arthritis (degenerative)	−.5
Atrophy	−.5
Bowel Obstruction	−.5
Cystic Diseases	−.5
Emaciation	−.5
Gout	−.5
Hydrocephalus (with air study)	−.5
Hyperparathyroidism	−.5
Leprosy	−.5
Multiple Myeloma	−.5
Necrosis	−.5
Osteomyelitis (active)	−.5
Pneumoperitoneum	−.5
Sarcoma	−.5
Syphilis (advanced)	−.5

TABLE M

Patient Age	= Bits
Birth to 3 months	= 1.6
3 months to 2 years	= 1.2
Extreme Age	= 1.0
2 years to 5 years	= .8
5 years to 12 years	= .4
Average Adult	= 0

TABLE N
GRID TRANSMISSION
(80 kVp)

Ratio	= Bits
None	= 10.0
4:1 60L	= 9.0
5:1 80L, 6:1 60L	= 8.4
8:1 80L, 10:1 103L	= 8.0
12:1 103L	= 7.6
5:1 Cross Hatched	= 7.4
16:1 80L	= 7.2
8:1 Cross Hatched	= 7.0

TABLE O
TUBE FILTRATION
(80 kVp)

Aluminum (mm Added)	= Bits
0	= 3.0
.5	= 2.8
1.0	= 2.6
1.5	= 2.4
2.0	= 2.2
3.0	= 2.0
4.0	= 1.8
5.0	= 1.6

TABLE P
CRONEX® FILM—SCREEN COMBINATIONS

Bit Speeds

Cronex® Screen		Cronex® Film			
	2 DC	4 & 6 +	6	7	Lo-dose
Quanta II =	17.2⁻	17.0	17.1	16.1	
Lightning Plus =	16.6	16.4	16.5	15.5	
Hi Plus =	16.2	16.0	16.1	15.1	
High Speed =	15.9	15.7	16.8	14.8	
Par Speed =	15.2	15.0	15.1	14.1	
Fast Detail =	14.3	14.1	14.2	13.2	
Lo-dose/2 (back only) =	14.0	13.8	13.9	12.9	12.0*
Lo-dose (back only) =	13.4	13.2	13.3	12.3	11.2*
Detail =	13.2	13.0	13.1	12.1	
No Screens =	9.8	9.6	9.7	8.4	8.6

Combinations with speeds below 14.0 Bits are recommended for thin parts (extremities) only.

* These speeds do not apply at 30 KvP and with beam filtration typical of mammography. Under mammography conditions Lo-dose/2 is two times the speed of Lo-dose.

TABLE Q
OPTIMUM DEVELOPMENT

Is Obtained When

Time Bits + Temp. Bits + Developer Activity Bits = 15.0

Developing Time* (Sec.) = Bits	TEMPERATURE (°F) = Bits = (°C)	Developer = Activity Bits
350 = 7.6	108 = 7.6 = 42.0	Cronex" XMD[1.] = 5.0
300 = 7.4	106 = 7.4 = 41.0	Cronex" XLD[2.] = 4.8
264 = 7.2	104 = 7.2 = 40.0	= 4.6
230 = 7.0	102 = 7.0 = 39.0	Cronex" XAD[3.] = 4.4
200 = 6.8	100 = 6.8 = 38.0	
167 = 6.6	98 = 6.6 = 37.0	
142 = 6.4	96 = 6.4 = 36.0	**USES**
120 = 6.2	94 = 6.2 = 34.5	
100 = 6.0	92 = 6.0 = 33.5	1. Automatic 90 sec & 3½ min.
86 = 5.8	90 = 5.8 = 32.0	2. Hand Tank
73 = 5.6	88 = 5.6 = 31.0	3. Automatic 3½ min. & 7 min.
61 = 5.4	86 = 5.4 = 30.0	
51 = 5.2	84 = 5.2 = 29.0	
43 = 5.0	82 = 5.0 = 28.0	
36 = 4.8	80 = 4.8 = 26.5	
30 = 4.6	78 = 4.6 = 25.5	
25 = 4.4	76 = 4.4 = 24.5	
21 = 4.2	74 = 4.2 = 23.5	**REPLENISHING RATES**
18 = 4.0	72 = 4.0 = 22.5	
15 = 3.8	70 = 3.8 = 21.0	Developer 70cc per 14 inch
13 = 3.6	68 = 3.6 = 20.0	Fixer 110cc per 14 inch
11 = 3.4	66 = 3.4 = 19.0	

*Immersion Time; no crossovers

TABLE R
ILLUMINATOR BRIGHTNESS

Fluorescent Tube		= Bits	=	Film
Foot Candles	Rel. EV			Base + Fog
175	11.5	= 5.6	=	.75
250	12.0	= 5.4	=	.60
350	12.5	= 5.2	=	.45
500*	13.0	= 5.0	=	.30
700	13.5	= 4.8	=	.15
1000	14.0	= 4.6	=	.00

*Recommended

TABLE S

Cones & Diaphrams	= Bits
8 x 10 or larger	= 5.0
5" x 7"	= 4.8
	4.6
Cylinder collapsed	= 4.4
Cylinder Extended	4.2

APPENDIX B. MILLIAMPERE-SECONDS FOR VARIOUS COMBINATIONS OF MILLIAMPERAGE AND EXPOSURE TIMES

Seconds	Milliamperes												
	25	50	100	150	200	300	400	500	600	700	800	1000	1200
0.003	0.075	0.150	0.3	0.45	0.6	0.9	1.2	1.5	1.8	2.1	2.4	3	3.6
.006	.150	.3	.6	.9	1.2	1.8	2.4	3	3.6	4.2	4.8	6	7.2
.009	.225	.45	.90	1.35	1.8	2.7	3.6	4.5	5.4	6.3	7.2	9	10.8
.012	.30	.60	1.2	1.80	2.4	3.6	4.8	6	7.2	8.4	9.6	12	14.4
.016	.40	.80	1.6	2.40	3.2	4.8	6.4	8	9.6	11.2	12.8	16	19.2
.020	.50	1	2	3	4	6	8	10	12	14	16	20	24
.025	.625	1.25	2.5	3.75	5	7.5	10	12.5	15	17.5	20	25	30
.032	.80	1.6	3.2	4.8	6.4	9.6	12.8	16	19.2	22.4	25.6	32	38.4
.040	1	2	4	6	8	12	16	20	24	28	32	40	48
.050	1.25	2.5	5	7.5	10	15	20	25	30	35	40	50	60
.064	1.60	3.2	6.4	9.6	12.8	19.2	25.6	32	38.4	44.8	51.2	64	76.8
.080	2	4	8	12	16	24	32	40	48	56	64	80	96
.10	2.5	5	10	15	20	30	40	50	60	70	80	100	120
.13	3.25	6.5	13	19.5	26	39	52	65	78	91	104	130	156
.16	4	8	16	24	32	48	64	80	96	112	128	160	192
.20	5	10	20	30	40	60	80	100	120	140	160	200	240
.25	6.25	12.5	25	37.5	50	75	100	125	150	175	200	250	300
.33	8.25	16.5	33	49.5	66	99	132	165	198	231	264	330	396
.7	17.5	35	70	105	140	210	280	350	420	490	560	700	840
1.0	25	50	100	150	200	300	400	500	600	700	800	1000	1200

APPENDIX C. DECIMAL EQUIVALENTS OF FRACTIONAL EXPOSURES

Denominators	Numerators											
	1	2	3	4	5	6	7	8	9	10	11	12
2	0.5	1.										
3	.333	.666	1.									
4	.25	.5	.75	1.								
5	.2	.4	.6	.8	1.							
6	.167	.333	.5	.667	.833	1.						
7	.143	.286	.429	.572	.715	.858	1.					
8	.125	.25	.375	.5	.625	.75	.875	1.				
9	.111	.222	.333	.444	.555	.666	.777	.888	1.			
10	.1	.2	.3	.4	.5	.6	.7	.8	.9	1.		
11	.09	.18	.27	.363	.455	.545	.636	.727	.818	.090	1.	
12	.083	.167	.25	.333	.415	.5	.583	.667	.75	.833	.917	1.
15	.067	.134	.2	.267	.333	.4	.467	.533	.6	.667	.733	.8
20	.05	.1	.15	.2	.25	.3	.35	.4	.45	.5	.55	.6
24	.042	.083	.125	.167	.208	.25	.292	.333	.375	.416	.458	.5
30	.033	.067	.1	.133	.167	.2	.233	.267	.3	.333	.367	.4
40	.025	.05	.075	.1	.125	.15	.175	.2	.225	.25	.275	.3
60	.017	.033	.05	.067	.083	.1	.117	.133	.15	.167	.183	.2
120	.008	.017	.025	.033	.042	.05	.058	.067	.075	.083	.092	.1

APPENDIX D. SQUARES, SQUARE ROOTS, AND RECIPROCALS OF SMALL NUMBERS

n	n^2	$\sqrt{n}$	$\frac{1}{n}$	n	n^2	$\sqrt{n}$	$\frac{1}{n}$
1	1	1	1	51	2601	7.14	.02
2	4	1.41	.5	52	2704	7.21	.019
3	9	1.73	.333	53	2809	7.28	.019
4	16	2	.25	54	2916	7.35	.019
5	25	2.24	.2	55	3025	7.42	.018
6	36	2.45	.166	56	3136	7.48	.018
7	49	2.64	.142	57	3249	7.55	.018
8	64	2.83	.125	58	3364	7.61	.017
9	81	3	.111	59	3481	7.68	.017
10	100	3.16	.1	60	3600	7.74	.017
11	121	3.32	.091	61	3721	7.81	.016
12	144	3.46	.083	62	3844	7.87	.016
13	169	3.60	.076	63	3969	7.94	.016
14	196	3.74	.07	64	4096	8	.016
15	225	3.87	.066	65	4225	8.06	.015
16	256	4	.063	66	4356	8.12	.015
17	289	4.12	.059	67	4489	8.18	.015
18	324	4.24	.055	68	4624	8.25	.015
19	361	4.36	.053	69	4761	8.31	.014
20	400	4.47	.05	70	4900	8.37	.014
21	441	4.58	.048	71	5041	8.43	.014
22	484	4.69	.045	72	5184	8.48	.014
23	529	4.79	.043	73	5329	8.54	.014
24	576	4.80	.042	74	5476	8.60	.014
25	625	5	.04	75	5629	8.66	.013
26	676	5.10	.038	76	5776	8.72	.013
27	729	5.10	.037	77	5929	8.77	.013
28	784	5.29	.036	78	6084	8.83	.013
29	841	5.38	.034	79	6241	8.89	.013
30	900	5.48	.033	80	6400	8.94	.013
31	961	5.57	.032	81	6561	9	.012
32	1024	3.66	.031	82	6724	9.05	.012
33	1089	5.74	.030	83	6889	9.11	.012
34	1156	5.83	.029	84	7056	9.16	.012
35	1225	5.92	.028	85	7225	9.22	.012
36	1296	6	.028	86	7396	9.27	.012
37	1369	6.08	.027	87	7569	9.33	.012
38	1444	6.16	.026	88	7744	9.38	.011
39	1521	6.24	.026	89	7921	9.43	.011
40	1600	6.32	.025	90	8100	9.49	.011
41	1681	6.40	.024	91	8281	9.54	.011
42	1764	6.48	.024	92	8464	9.59	.011
43	1849	6.56	.023	93	8649	9.64	.011
44	1936	6.63	.023	94	8836	9.69	.011
45	2025	6.71	.022	95	9025	9.75	.011
46	2116	6.78	.022	96	9216	9.70	.01
47	2209	6.85	.021	97	9409	9.85	.01
48	2304	6.93	.021	98	9604	9.80	.01
49	2401	7	.02	99	9801	9.95	.01
50	2500	7.07	.02	100	10000	10	.01

APPENDIX E. ANSWERS TO REVIEW QUESTIONS.

CHAPTER 1

1. a. 0.5
   b. 0.033
   c. 0.016
   d. 0.008
   e. 0.100
2. a. 8.3 ms
   b. 16.66 ms
3. a. 2.77 ms
   b. 5.55 ms
4. a. 1.38 ms
   b. 2.77 ms
5. a. 55.88 cm (2.54 cm/in. × 22 in.)
   b. 182.88 cm (2.54 cm/in. × 72 in.)
   c. 91.44 cm (2.54 cm/in. × 36 in.)
   d. 2.27 kg (454 gm/lb × 5 lb)
   e. 180 ml
   f. 35°C
   g. 20°C
6. a. $mA_2 = 300$
   b. $t_2 = 200$ ms

7. a. $mA_2 = 400$
   b. $d_2 = 70.7$ cm
8. Solution:

$$\frac{R_1}{R_2} = \left(\frac{d_2}{d_1}\right)^2$$

$$R_1 = 5mR$$

$$R_2 = x$$

Hence: $\dfrac{5}{x} = \left(\dfrac{80}{100}\right)^2 = \left(\dfrac{4}{5}\right)^2$

$$\frac{5}{x} = \frac{16}{25}$$

$$16x = 125; \ x = 7.8$$

9. A phototimer terminates the radiographic exposure at the time necessary to obtain a correct film density. Selecting a higher mA station will normally shorten the exposure time.

CHAPTER 2

1. e
2. b
3. b
4. b
5. e
6. c
7. d
8. a
9. b
10. a
11. faster, slower
12. e
13. a
14. a
15. a
16. b
17. d
18. c
19. b
20. b
21. d
22. d
23. b
24. a
25. c

CHAPTER 3

1. b
2. d
3. b
4. b
5. c
6. c
7. c
8. c
9. a
10. e
11. e
12. d
13. d
14. d
15. c
16. e
17. d
18. b
19. b
20. d

CHAPTER 4

1. c	8. c	15. b
2. d	9. d	16. c
3. b	10. a	17. e
4. d	11. c	18. d
5. c	12. c	19. c
6. b	13. e	20. d
7. b	14. a	

CHAPTER 5

1. a	9. e	17. a
2. a	10. b	18. a
3. c	11. b	19. b
4. e	12. c	20. a
5. e	13. c	21. b
6. b	14. b	22. b
7. d	15. b	23. b
8. a	16. c	24. a

CHAPTER 7

1. e	11. b	21. d
2. e	12. d	22. PE = 10, T = 1
3. e	13. b	23. PE = 49, mA = 200
4. a	14. b	24. PE = 49, mA = 125
5. b	15. b	25. PE = 122.5, SID = 44.7
6. b	16. c	26. b
7. a	17. a	27. b
8. b	18. a	28. a
9. b	19. a	29. b
10. a	20. a	30. b

CHAPTER 8

1. c	3. 8.5 mAs	7. a
2. New mAs = 10	4. a	8. c
New mAs = 1.66	5. c	
New kV = 98	6. c	

CHAPTER 10

1. e	7. c	13. a
2. a	8. b	14. b
3. c	9. b	15. c
4. b	10. b	16. a
5. c	11. b	
6. c	12. b	

CHAPTER 11

1. b	10. d	19. b
2. b	11. c	20. a
3. d	12. e	21. a
4. b	13. c	22. b
5. a	14. a	23. e
6. b	15. c	24. a,b
7. d	16. c	25. a
8. a	17. c	
9. a	18. a	

APPENDIX F. NATIONAL BOARD EXAMINATIONS: TYPICAL QUESTIONS

PREPARING FOR NATIONAL BOARD EXAMINATIONS

Many students who must take national certifying examinations have, unfortunately, not had experience in taking Board type examinations. Board examinations are different from most examinations taken by radiologic technology students during their educational training. Included here are Board examination questions typical in variety and similar to those used on certifying examinations. The student should become familiar with the type and nature of examination questions used. The precise content of most of the questions included here probably would not be duplicated on a Board examination.

There are three basic points that any student should remember in preparing for Board examinations. First, most examinations are designed to test knowledge of the commonplace or the basic experience rather than the esoteric. In preparing for the examination make sure you know the essential, even if obvious, information; spend most of your time reviewing basic data, not the rarely used data. Second, read the instructions and do exactly as the instructions indicate. Third, don't change an answer unless you are absolutely sure that you have marked it wrong. Often an answer is changed from right to wrong.

SAMPLE EXAMINATION FOR RADIOLOGIC TECHNOLOGISTS

Directions: Each of the questions or incomplete statements below is followed by five suggested answers or completions. Select the one which is best in each case.

1. The apices of the lung are best demonstrated by

   a. Apical lordotic view
   b. Lateral view of the chest
   c. PA view of the chest
   d. Right posterior oblique
   e. Right lateral decubitus

2. A minute-sequence urogram is primarily used in evaluation of

   a. Renal stones
   b. Renovascular hypertension
   c. Pyelonephritis
   d. Perinephric abscess
   e. Renal cell carcinoma

3. Which of the body tissues listed below will absorb the LEAST radiation?

   a. Muscle
   b. Bone
   c. Fat
   d. Blood
   e. Nerves

4. Which of the following is located in the right upper quadrant of the abdomen?

   a. Spleen
   b. Appendix
   c. Cecum
   d. Liver
   e. Jejunum

5. The x-ray technician's primary responsibility is to the

    a. Radiologist
    b. Referring physician
    c. Supervisor
    d. Hospital
    e. Patient

6. In a hospital setting, film obtained from an x-ray examination legally belongs to the

    a. Radiologist
    b. Referring physician
    c. Hospital
    d. Patient
    e. Chief technician

7. Rotation of the body in positioning for a radiograph is primarily to overcome

    a. Magnification
    b. Motion
    c. Distortion
    d. Superimposition
    e. All of the above

8. Which of the following is (are) *not* normally exposed to radiation on a transverse 14" x 17" film of the pelvis?

    a. Ovaries
    b. Adrenal glands
    c. Rectum
    d. Anus
    e. Sigmoid colon

9. The colon contains all of the following portions *except*

    a. Cecum
    b. Sigmoid
    c. Rectum
    d. Splenic flexure
    e. Ileum

10. A 28-year-old radiologic technologist has been exposed to radiation for a period of 7 years. His maximum accumulation of whole body radiation is

    a. 35 rem
    b. 50 rem
    c. 110 rem
    d. 100 rem
    e. 56 rem

11. For the average person and for radiation safety purposes, the end of the childbearing age is

    a. 25
    b. 28
    c. 30
    d. 32
    e. 35

12. In determining shielding for radiation safety, all of the following are considered *except*

    a. Distance
    b. Occupancy rate
    c. Use factor
    d. Work load
    e. Age of employee

13. Ways of reducing radiation exposure to the patient include which of the following?

    I. X-ray beam collimated to size of film being used
    II. Use of high speed screens
    III. Use of high kilovoltage techniques

    a. I only
    b. II only
    c. I and II only
    d. II and III only
    e. I, II, and III

14. Proper positioning of a patient for examination of the kidneys, ureter, and bladder include which of the following?

   I. Symphysis pubis
   II. Iliac crest
   III. Xiphoid process
   IV. Fifth lumbar vertebra

   a. I and III only
   b. II and IV only
   c. I, II, and III only

   d. II, III, and IV only
   e. I, II, III, and IV

15. What is the proper sequence for processing a radiographic film in an automatic processor?

   I. Fix
   II. Wash
   III. Develop
   IV. Dry

   a. I, II, III, IV
   b. III, I, IV, II
   c. III, IV, II, I

   d. II, III, I, II, IV
   e. III, I, II, IV

Directions: Each group of questions below concerns a certain situation. In each case, first study the description of the situation. Then choose the one best answer to each question following it.

Questions 16–20

A 65-year-old patient was sent to the x-ray department for an intravenous cholangiogram. During injection of the contrast media, the patient became anxious, began perspiring, and had difficulty in breathing. A few minutes later he was unconscious and his pulse was barely palpable.

16. The most logical impression is that this patient

   a. Had a heart attack
   b. Threw a blood clot to the lungs
   c. Had an adverse reaction to contrast media
   d. All of the above
   e. None of the above

17. The first step one should take in care of the patient is to

   a. Establish an adequate airway
   b. Run for help
   c. Hook up the EKG machine
   d. Pound on his chest
   e. Take an x-ray film before contrast media is gone

18. Assuming that the contrast media was injected using the drip-infusion method, which of the following is most appropriate?

   a. Continue infusion of the contrast media so as not to ruin the examination
   b. Remove the infusion needle and apply pressure on the vein
   c. Remove the infusion set but leave the needle in place
   d. Turn off infusion of contrast media leaving the needle in place
   e. Ask the supervisor what to do

19. Before intravenous injection of iodinated contrast media, all of the following must be on hand *except*

    a. Antihistamine
    b. Adrenaline (Epinephrine)
    c. Steroids for injection
    d. Oxygen tank
    e. Cardiac drugs

20. Within the x-ray department in cardiac-pulmonary resuscitation, *initial* emergency care is the responsibility of the

    a. Patient's physician
    b. The radiologist
    c. X-ray supervisor
    d. Radiologic technologist
    e. Emergency team

Questions 21–24

A patient was sent to the x-ray suite with the following x-ray request:

X-ray Request			
Patient Identification:	*Age:*	*Examination Requested:*	*Location:*
Johnson, Ruby L.	55	AP, Lat, Oblique, C. Spine	E R
#489–54–32–17	*Sex:*	PA, Lat, Pelvis	*Transportation:*
	F		Stretcher
Pertinent Clinical History:			
Pregnant female in car accident. R/O fx., C and LS Spine.			

21. Assume that you are an emergency room x-ray technician with no one to help. Which of the following moves is the most appropriate?

    a. Ask for assistance in moving the patient from the stretcher to the x-ray table
    b. Help the patient from the stretcher to the x-ray table
    c. Take portable films
    d. Send the patient back to the ER since directions are unclear
    e. Leave the patient on the stretcher and use fixed radiographic equipment for the examination

22. The patient is conscious and identifies herself as "Ruby Bonson." Which of the following is the proper action?

    a. Assume that the x-ray requisition is in error and change the name on the request
    b. Call the emergency room and determine whether the patient's identity is correct
    c. Ignore the patient; take appropriate x-rays and appropriately identify the films
    d. Leave the patient identification blank to be filled in later
    e. Take appropriate x-rays and change the film identification if the patient's name is wrong

23. The cervical spine is x-rayed first. Which of the following is the proper procedure?

    a. AP and lateral films are taken first, and permission is asked to reposition the patient so that other views are obtained
    b. All views are taken and shown to the physician before the patient is moved or repositioned
    c. The obliques are taken first so as to demonstrate a pars interarticularis fracture
    d. A lateral film is taken and permission is requested from the physician before other films are made
    e. Permission to move the patient is obtained before each view is made

24. Directions for proper positioning technique for exposing the lateral cervical spine include all of the following *except*

    a. Use a vertical x-ray beam
    b. Center the central ray at C-5
    c. Lower both shoulders, with traction if necessary
    d. Immobilize standard grid cassette against shoulder parallel to neck and perpendicular with the central ray
    e. Technique of 70–75 kVp, 70 mAs, 72" FFD

Questions 25–31

A 43-year-old, average-sized patient is sent to the x-ray department for an IVP. His x-ray request and medical records indicate that he has viral hepatitis.

25. Hepatitis is usually contracted from a patient such as this by

    a. Blood and/or stool contamination of an open wound
    b. Inhalation of viral particles
    c. Venipunctures using sterile technique
    d. All of the above
    e. None of the above

26. Jaundice which usually accompanies hepatitis may be recognized by

    a. Yellow, sallow appearance of patient
    b. Profuse perspiration
    c. Weak, anemic appearance of patient
    d. All of the above
    e. A and C only

27. The proper procedure for disposing of linen, syringes, and needles is

    a. All should be isolated, properly labeled, and disposed of according to your hospital regulations
    b. Linen may be placed with other dirty linen from the department
    c. Needles are isolated and disposed of separately with linen and syringes handled routinely
    d. No special precautions are needed
    e. The IVP should be cancelled because of the contagious nature of hepatitis

28. An oblique film to demonstrate the left kidney is made. Which of the following is the correct position?

    a. LPO, 60°   d. RAO, 20°
    b. LPO, 20°   e. RAO, 60°
    c. RPO, 60°

29. The exposure factors for the KUB on the above patient was 70 kVp, 200 mAs, single-phase equipment. The exposure factors for a coned oblique of the left kidney would be approximately

    a. 70 kVp, 200 mAs   d. 80 kVp, 200 mAs
    b. 70 kVp, 100 mAs   e. 80 kVp, 400 mAs
    c. 74 kVp, 200 mAs

30. The x-ray requisition asked for a PA and lateral pelvis examination; however, the medical history suggests a lumbosacral spine fracture. Which of the following is most appropriate?

    a. The films should be taken as requested, with the abdomen shielded

b. Disregard the abdominal shielding since this may obscure a fracture

c. Obtain a lumbosacral spine series since the history suggests a fracture of the spine rather than the pelvis

d. Refuse to do the examination since the patient is pregnant

e. Ask the radiologist or referring physician to clarify the examination

31. In obtaining a lateral film of the lumbosacral spine of an average patient which of the following is *not* true?

    a. Vertical x-ray beam

    b. Patient in true lateral position with knees flexed

    c. X-ray beam centered just above the iliac crest

    d. Patient takes a long, slow breath while exposure is being made

    e. Technical factors of 85–90 kVp, 200 mAs, 40" FFD.

Directions: Each group of questions below consists of five lettered headings followed by a list of numbered words, phrases, or sentences. For each numbered word, phrase, or sentence select the one heading which is most closely related to it. One heading may be used once, more than once, or not at all in each group.

Questions 32–37

    a. Barium sulfate USP

    b. Telepaque (Iopanoic acid USP)

    c. Renografin-60 (Meglumine Diatrizoate USP)

    d. Dionosil Oily (Propliodone oil suspension USP)

    e. Pantopaque (Iophendylate USP)

32. Myelography

33. Upper gastrointestinal series

34. Gallbladder series

35. Intravenous urograms (IVP)

36. Bronchography

37. Angiography

Questions 38–43

    a. Knee

    b. Shoulder

    c. Ankle

    d. Hip

    e. Foot

38. Lateral malleolus

39. Patella

40. Talus

41. Glenoid fossa

42. Greater trochanter

43. Metatarsal

Cahoon's Formulating X-Ray Techniques

Questions 44–49

Proper automatic processing of medical x-ray film is an important factor in maintaining quality control in an x-ray department. For common processing problems listed in questions 44–49, select from a-e the most appropriate response.

a. Film fogged
b. Static marks on film
c. Tacky film
d. Light films coming from processor
e. Films jamming at fixer rack assembly

44. Processor developer temperature too high

45. Processor developer temperature too low

46. Inadequate fixer replenishing rate

47. Humidity too low in darkroom

48. Broken spring in crossover rack

49. Inadequate developer replenishing rate

Questions 50–55

On occasion films may be ruined because of inadequate radiographic technique not related to exposure factors. For questions 50–55, choose the most appropriate answer.

a. Poor film-screen contact
b. Pitted or dirty intensifying screens
c. Light leak from edge of cassette
d. Improper use of collimator or Bucky tray
e. Cassette or film exposed to scattered radiation

50. Film fogged

51. Black mottling on edge of film

52. Grid marks on radiograph

53. Localized blurring on radiograph

54. Persistent small white defects on radiograph

55. Clear edge to one side of film

Questions 56–61

Assume single-phase equipment, an average adult patient, and appropriate FFD. What is the average technique for the following examinations?

a. 70 kVp, 10 mAs
b. 70 kVp, 80 mAs
c. 70 kVp, 200 mAs
d. 120 kVp, 3 mAs
e. 120 kVP, 50 mAs

56. Lateral skull

57. PA chest

58. AP abdomen, IVP

59. AP abdomen, barium enema

60. True lateral hip

61. AP clavicle

Directions: For each of the questions below, *one* or *more* of the responses given are correct. Decide which of the responses is (are) correct.

    a. if *1, 2, and 3* are correct;
    b. if only *1 and 2* are correct;
    c. if only *2 and 3* are correct;
    d. if only *1* is correct;
    e. if only *3* is correct.

Questions 62–70

62. Kilovoltage controls

    1. Density
    2. Contrast
    3. Penetration

63. Emphysema results in

    1. Increased lung volume
    2. Decreased air exchange
    3. Increased penetration required

64. High kV techniques are useful in

    1. Extremity radiography
    2. Abdominal films in pregnancy
    3. Lateral films of lumbar spine

65. Static electricity in the darkroom can be controlled by

    1. Increasing humidity
    2. Wearing rubber gloves
    3. Increasing development temperature

66. The reduction in radiation film fog by use of a grid depends on

    1. Grid ratio
    2. Lines per inch
    3. Strip density

67. Focal spot size of an x-ray tube is dependent on

    1. Milliamperage applied to cathode
    2. Kilovoltage applied across the tube
    3. Focal film distance

68. Contrast agents used in myelography include

    1. Metrizamide
    2. Pantopaque
    3. Air

69. At 183 cm. SID, a focal spot must be

    1. Less than 1 millimeter
    2. Greater than 2 millimeters
    3. Any size, within reason

70. Standard views for evaluation of a possible fracture in a nine-year-old include
    1. AP
    2. Lateral
    3. Opposite view

Directions: Each group of questions below refers to a diagram or radiograph with certain parts labeled with numbers. Each question is followed by five suggested labels or answers. For each question select the one best answer.

Questions 71–73

Refer to the accompanying 3-phase tube-rating chart.

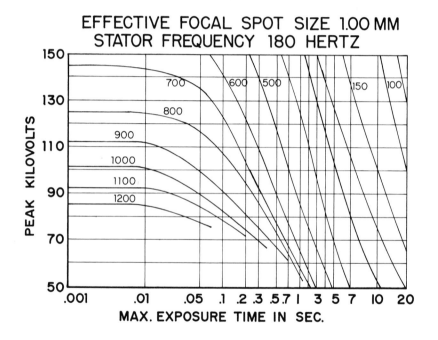

EFFECTIVE FOCAL SPOT SIZE 1.00 MM
STATOR FREQUENCY 180 HERTZ

71. The kilowatt rating of this x-ray tube is approximately

    a. 50   d. 100
    b. 75   e. 150
    c. 85

72. What is the maximum exposure time that can be made at 90 kV, 900 mA station?

    a. .01   d. .2
    b. .05   e. .5
    c. .10

73. At 80 kV, .01 sec. exposure, what is the largest mA station that may be used?

    a. 800   d. 1100
    b. 900   e. 1200
    c. 1000

Questions 74–78

Refer to the accompanying labeled diagram of the heart.

PA PROJECTION

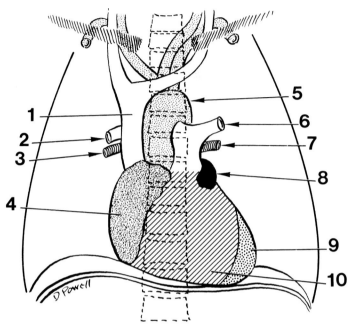

74. The left atrial appendage is structure
    a. 4   b. 6   c. 8   d. 9   e. 10

75. The right ventricle is structure
    a. 4   b. 6   c. 8   d. 9   e. 10

76. The superior vena cava is structure
    a. 1   b. 2   c. 3   d. 5   e. 7

77. The pulmonary artery is structure
    a. 1   b. 3   c. 5   d. 6   e. 7

78. The left ventricle is structure
    a. 4   b. 5   c. 8   d. 9   e. 10

Directions: Each of the questions or incomplete statements below is followed by five suggested answers or completions. Select the one which is best in each case.

79. The right major fissue separates
    a. Superior from inferior lingula
    b. Middle lobe from the upper and lower lobes
    c. Anterior segment of the upper lobe from posterior segment
    d. Upper and middle lobes from the lower lobe
    e. Posterior basal segment from superior basal segment

80. The *primary* use of lymphangiography is to

    a. Evaluate lymphedema
    b. Stage lymphoma
    c. Evaluate peritoneal granulomatous disease
    d. Evaluate varicose veins
    e. Evaluate inflammatory lesions of the feet

81. Which of the below examinations demonstrate *both* vertebral and carotid circulation?

    a. Right brachial arteriogram
    b. Right carotid arteriogram
    c. Left brachial arteriogram
    d. All of the above
    e. Both a and c

82. Which of the below is a midline cerebral vessel?

    a. Right carotid artery
    b. Anterior cerebral artery
    c. Vertebral artery
    d. Middle cerebral artery
    e. Posterior cerebellar artery

83. Which of the body tissues listed below will absorb the *most* radiation?

    a. Fat
    b. Muscle
    c. Blood
    d. Bone
    e. Air

84. Which of the body tissues listed below will absorb the *least* radiation?

    a. Fat
    b. Muscle
    c. Nerves
    d. Bone
    e. Air

85. A fluoroscopic examination is used primarily to evaluate

    a. Motion
    b. Children
    c. Roentgen anatomy
    d. Positioning of the patient
    e. Chest pathology

86. In an x-ray tube, x-rays are produced in the

    a. Anode
    b. Cathode
    c. Filament
    d. Penumbra
    e. Focusing cup

87. Matter which is penetrated by x-rays with relative ease is termed

    a. Radiopaque
    b. Radiolucent
    c. Radiopathology
    d. Photolucent
    e. Photosensitive

88. In protecting oneself from exposure to sources of radiation, which of the following factors must be considered?

    a. Distance
    b. Shielding
    c. Time
    d. All of the above
    e. None of the above

89. Matter which absorbs x-rays is termed

    a. Radiopaque
    b. Radiolucent
    c. Radiopathology
    d. Photolucent
    e. Photosensitive

90. A minute sequence urogram is primarily used in evaluation of

    a. Renal stones
    b. Renovascular hypertension
    c. Pyelonephritis
    d. Perinephritic abscess
    e. Renal cell carcinomas

91. An expiratory/inspiratory film taken during an IVP sequence is useful in

    a. Measuring movement of diaphragms
    b. Determining whether gallstones are present
    c. Determining movement of kidneys
    d. Determining displacement of ureters
    e. Isolating renal from adrenal masses

92. Radiographic contrast is primarily influenced by

    a. Milliamperes-seconds
    b. Exposure time
    c. Milliamperage
    d. Kilovoltage
    e. Source/image-receptor distance

93. When using three-phase radiographic equipment, heat units are calculated by

    a. mA × time × 0.7
    b. kV × mAs × 1.4
    c. kV × mA × time
    d. kV × mA × SID
    e. kV × mAs × 0.7

94. With the patient in the supine position, the orbital-meatal line is placed parallel to the film. This position is called

    a. Posteroanterior position
    b. Modified Towne position
    c. Submentovertex position
    d. Towne position
    e. Rhumstead II position

95. A shallow breathing technique is useful for evaluation of

    a. Neck
    b. Kidneys
    c. Sternum
    d. Lung fields
    e. Gastrointestinal tract

96. In an AP radiograph of the sacrum, the x-ray beam is directed

    a. Perpendicular to the film
    b. Cephalad
    c. Caudad
    d. Towards S 4
    e. Lateral to L 3

97. In 12-pulse three-phase x-ray equipment, how many pulses would be seen on a spin top for a 100 ms exposure?

    a. None
    b. 30
    c. 60
    d. 90
    e. 120

Cahoon's Formulating X-Ray Techniques

Directions: Given the exposure factors listed below, choose the best combinations for questions 98–100. Each answer may be used once, more than once, or not at all.

	mA	kV	Time	SID
a.	100	65	100 ms	90 cm
b.	50	75	200 ms	90 cm
c.	1000	75	10 ms	183 cm
d.	500	90	50 ms	90 cm
e.	300	80	80 ms	150 cm

98. Film with greatest contrast

99. Shortest exposure time

100. Least magnification distortion

ANSWERS TO NATIONAL BOARD EXAMINATIONS: TYPICAL QUESTIONS

1. a	26. a	51. c	76. a
2. b	27. a	52. d	77. d
3. c	28. b	53. a	78. d
4. d	29. c	54. b	79. d
5. e	30. e	55. d	80. b
6. c	31. d	56. b	81. a
7. d	32. e	57. d	82. b
8. b	33. a	58. c	83. d
9. e	34. b	59. e	84. e
10. b	35. c	60. b	85. a
11. d	36. d	61. a	86. a
12. e	37. c	62. a	87. b
13. e	38. c	63. b	88. d
14. c	39. a	64. c	89. a
15. e	40. e	65. d	90. b
16. c	41. b	66. a	91. c
17. a	42. d	67. b	92. d
18. d	43. e	68. a	93. b
19. d	44. a	69. e	94. a
20. d	45. d	70. a	95. c
21. a	46. c	71. c	96. b
22. b	47. b	72. c	97. a
23. d	48. e	73. c	98. a
24. a	49. d	74. c	99. c
25. a	50. e	75. e	100. c

INDEX

Index

Index

Index

Index